# THE UNITED STATES AND MULTILATERAL RESOURCE MANAGEMENT

# THE UNITED STATES AND MULTILATERAL RESOURCE MANAGEMENT

Edited by
Robert S. Jordan

PRAEGER

PRAEGER SPECIAL STUDIES • PRAEGER SCIENTIFIC

New York • Philadelphia • Eastbourne, UK
Toronto • Hong Kong • Tokyo • Sydney

**Library of Congress Cataloging in Publication Data**
Main entry under title:

The United States and multilateral resource management.

Bibliography: p.
Includes index.
1. Natural resources--International cooperation--Addresses, essays, lectures. 2. Marine resources--International cooperation--Addresses, essays, lectures. 3. Power resources--International cooperation--Addresses, essays, lectures. 4. Agriculture--International cooperation--Addresses, essays, lectures. 5. United States--Foreign economic relations--Addresses, essays, lectures. I. Jordan, Robert S., 1929–
HC59.U54 1985 333.7'0973 85-6321
ISBN 0-03-060311-0 (alk. paper)

Published in 1985 by Praeger Publishers
CBS Educational and Professional Publishing, a Division of CBS Inc.
521 Fifth Avenue, New York, NY 10175 USA



56789 052 987654321

Printed in the United States of America on acid-free paper

## INTERNATIONAL OFFICES

Orders from outside the United States should be sent to the appropriate address listed below. Orders from areas not listed below should be placed through CBS International Publishing, 383 Madison Ave., New York, NY 10175 USA

**Australia, New Zealand**
Holt Saunders, Pty, Ltd., 9 Waltham St., Artarmon, N.S.W. 2064, Sydney, Australia

**Canada**
Holt, Rinehart & Winston of Canada, 55 Horner Ave., Toronto, Ontario, Canada M8Z 4X6

**Europe, the Middle East, & Africa**
Holt Saunders, Ltd., 1 St. Anne's Road, Eastbourne, East Sussex, England BN21 3UN

**Japan**
Holt Saunders, Ltd., Ichibancho Central Building, 22-1 Ichibancho, 3rd Floor, Chiyodaku, Tokyo, Japan

**Hong Kong, Southeast Asia**
Holt Saunders Asia, Ltd., 10 Fl, Intercontinental Plaza, 94 Granville Road, Tsim Sha Tsui East, Kowloon, Hong Kong

**Manuscript submissions should be sent to the Editorial Director, Praeger Publishers, 521 Fifth Avenue, New York, NY 10175 USA**

When institutions do not behave the way they are expected to, men may ultimately have to change their ideas. In the short run, however, a typical human response is the development of fictions that serve to obscure the contradiction between expectation and reality.

-- Andrew M. Scott

This book is dedicated to:

Dr. Martin Hickman

As Dean of the College of Family, Home and Social Sciences of Brigham Young University, Dean Hickman has brought to the university a strong commitment to the study of the international dimension of national life and has supported research to this end. He has also had a lasting concern for the moral and ethical aspects of international affairs.

# ACKNOWLEDGMENTS

The idea for a book on this subject arose while I was serving as Director of Research at the United Nations Institute for Training and Research (UNITAR). In this capacity, I was responsible for a series of policy studies dealing with the so-called "sectoral" issues taken up in the ad hoc global conferences of the 1970s--food, energy, habitat, population, water, new and renewable resources, exploration and exploitation of the sea-bed, science and technology for development, etc. It became clear to me that there has been a fundamental clash not only in the substantive issues and negotiating techniques, but also in the values between the various blocs of states as they formed coalitions over these issues. Then, while serving as the first Dag Hammarskjold Visiting Professor of International Relations at the University of South Carolina in 1979-81, I had occasion to participate in a policy study on the impact of the Law of the Sea negotiations on United States interests concerning offshore marine mineral exploration and exploitation. Dr. James Kuhlman, then Chairman of the Department of Government and International Studies, was instrumental in making possible this opportunity.

From these two experiences, I organized and conducted a conference on the subject, "The United States and Multilateral Resource Management," at Brigham Young University in May 1984, under a grant from the University's David M. Kennedy Center for International Studies. The draft chapters of this book provided the framework for the discussion, and the Introduction in this book is a report on the conference itself. Participants were: David Adamson, Joseph P. Black, Roger Coate, Robert Crawford, Robert Friedheim, Christopher Herrick, Martin J. Hillenbrand, Constance Lundberg, Gerard Mangone, Donald Puchala, John P. Renninger, Robert

Riggs, Louis J. Royer, Burton Sapin, Janet Schmidt, James Sutterlin, Joseph Szylowicz, Dennis Thompson.

I am especially grateful to the Faculty Support Center of the College of Family, Home and Social Sciences of Brigham Young University for assistance in preparing the manuscript for publication, to Dr. Stan Taylor, Director of the David M. Kennedy International Center, to his colleagues Dr. Spencer Palmer and Dr. Ray Hillam, and to members of the Center's staff for assistance in solving the many logistical problems associated with convening a conference of this kind, and to Louis Royer who served ably as my research assistant.

Dr. Robert S. Wood, Director of the Center for Naval Warfare Studies, Naval War College, provided encouragement, and the Word Processing Center of the College provided essential services in preparing final camera-ready copy.

Robert S. Jordan
Distinguished Visiting Professor
Department of Policy and Strategy
Naval War College (on leave from the University of New Orleans)

January 1985

# CONTENTS

# INTRODUCTION: THE UNITED STATES AND MULTILATERAL RESOURCE MANAGEMENT: ISSUES IN ENERGY, FOOD, AND MARINE MINERALS A CONFERENCE REPORT

*John P. Renninger*

The "national interest," or more precisely the delineation of the United States national interest, is a source of continuing debate and discussion in both academic and political circles. And in an increasingly interdependent world, the identification of the "national interest" becomes an even more hazardous undertaking. It is clear, for example, that in some areas of national life the national interest can best be promoted through the use of multilateral institutions, such as military alliances. But there is no clear overall understanding of the advantages of multilateralism as opposed to unilateralism or bilateralism, or an appreciation of the trade-offs among them that are inevitably involved.

This conference dealt with these questions with regard to three resource issues: marine minerals, energy, and food. The goal of the conference, sponsored by the David Kennedy Center of Brigham Young University, was to discuss the trade-offs in each of these three issue areas.

In the discussion of marine minerals, the Chair (Gerard Mangone) noted the importance of minerals in history (bronze age, iron age, etc.). Until the 1950s there was, however, not much interest in marine minerals. Then expectations of huge wealth in the seabed led to the Conference on the Law of the Sea, and ultimately to the Treaty on the Law of the Sea. The minerals in the seabed were related to the notion of common property rights, which has become a central global problem. The

Treaty contains a framework for the management of the "commons" of the seabed, but the United States has not signed the Treaty, and prospects for its implementation are cloudy.

Much of the discussion on this issue therefore concerned the hope and prospects for a global commons approach to the seabed, although a specific issue of concern was the Treaty itself and the decision of the United States not to sign it. The attributes of an ocean commons (physical entity that is difficult to divide, useful and open) were noted. As a way of dealing with scarcity, the ocean commons are today being reduced by various forms of enclosure, and a frequent comment in the discussion was that in the long run this could lead to disaster.

It was asserted and generally accepted that the availability of marine minerals did not, in fact, constitute a short-term problem since there was no scarcity of them. The consequences of scarcity therefore were not vital for either the developed or developing countries. The politics surrounding the negotiations therefore have been as much symbolic as substantive: little real cost has been involved for either side. Contrary to earlier expectations, it is now generally recognized that the effort involved in exploiting marine minerals would likely produce no bonanza for the American and other Western companies that would be in a position to engage in such exploitation. The proposed International Seabed Authority was therefore unlikely to be the beneficiary of great revenues which could be used for projects favored by the Third World. Furthermore, given the costs of marine exploitation, land-based producers of the same minerals do not feel overly threatened.

In the debate over the Law of the Sea Treaty, then, vital national economic interests have not been involved. However, the freedom of passage through straits aspect of the Treaty does involve strategic national interests. But here the United States view (perhaps excessively optimistic) has been that international legal principles and conventions already protect American interests.

It was generally agreed that the major resistance to the Treaty came from private corporations which did not see their corporate interests served by United States involvement in the International Seabed Authority and Enterprise. Their misgivings coincided with the support of the Reagan Administration for private rather than public enterprise. The present general disillusionment with using international organizations to advance national interests was also a factor. Even though the Treaty was a package deal and contained many elements sought by the United States (passage through straits, etc.), the ideological dimension and the access of the multinationals to policy makers in the United States government proved stronger. However, some of the participants noted that it would be a mistake to identify the United States national interest too closely with those of the multinationals, many of whom were moving their operations out of the United States and whose interests were becoming separate from those of the United States. In cases such as these, the national interest is particularly difficult to discern.

It was also felt to be an illusion to believe that the United States could in any case retain control over the seabed. The issue was not control but interest, and in the long term a parochial approach would not serve the national interest.

Many of the participants felt that, in any event, the International Seabed Authority and the Enterprise would not be very efficient multilateral mechanisms for dealing with marine minerals in a global commons approach. This explains part of the United States resistance. It was also explained at length that the United States felt, perhaps with justification, that it could reap the advantages of the Treaty without being a party to it. Yet it was, at the same time, agreed that efficiency was not the only value to be considered. Others might be security and equity. While, as one participant noted, it would require a "leap of faith" to believe that the institutions created by the Treaty would work, many felt that an opportunity had been lost because the Treaty did represent an attempt to deal with a problem within a global commons framework before exploitation started. Such desired goals of multilateralism as resource conservation, environmental protection, and orderly development could perhaps be served by the Treaty. The Treaty also had very important implications for the future development of the concept of international organization and the progressive development of international law. Even though it was recognized that technological developments already had made much of the Treaty obsolete, it could at least provide an appropriate framework for future negotiations.

The participants were in agreement that it was unlikely that warfare would develop over marine minerals, at least in the near future. However, the question was raised as to what would happen if United States-sponsored corporations tried to exploit unilaterally marine minerals in the seabed. Would a case be brought against them or their governments in the World Court? But a more vital question concerned resolving disputes among the developed countries capable of exploiting the minerals in the seabed. A framework for resolving such potential conflicts was required which could perhaps form an essential ingredient in an eventual international regime.

The second issue considered by the conference was energy. The Chair (Martin Hillenbrand) led off the discussion by asserting that the United States government was now quite "relaxed" about the whole energy issue, although the economic consequences of the two oil shocks were still only dimly appreciated. There had been radical changes in the structure of the world oil market which were still going on. For example, in addition to sharp price fluctuations, immense substitution was also now occurring. There would be no pressing worries about supplies of oil for the remainder of the century. However, it was possible that in the late 1980s or 1990s there could be upward price pressure. This at least was the opinion of the International Energy Agency, which had been underrated by the United States and was certainly not a failed international organization. Despite this danger, the United States government, overtaken by lethargy and indifference, has not created any institutional machinery for dealing with a domestic shortage. This is true despite the fact that the United States was still bound by agreements for the pooling of resources, etc.

In the discussion, there was some debate as to whether oil (and energy in general) should be considered a global commons issue. Some participants felt that there was no identifiable global commons issue except potentially over the oil in the continental shelf. Other participants were of the opinion that the global commons concept need not be limited to the sea, the atmosphere, etc., but should involve access to a resource so universally consumed. These participants were of the opinion that the whole question of energy and oil could be intimately related to certain other New International Economic Order (NIEO) issues. Other participants doubted the sincerity of the OPEC states in insisting on this linkage in the global negotiations, but the possible relevance of energy to the larger issue of global resource redistribution was not disputed.

It was agreed that at present there are two international organizations representing, respectively, the interests of the main consumer states and the main producer states (International Energy Agency and OPEC). A major question which was discussed at length concerned the need or desirability of additional international mechanisms, more global and universal in nature. Some participants felt that with the likelihood of shortages at the end of the decade, additional multilateral action would be desirable. The problems that could perhaps be solved by a multilateral approach included those related to information, dissemination, production, and defense of sources of supply.

But the United Nations at the present time has no agency to deal with energy in a comprehensive manner. Such an agency might be able to deal with some of the problems mentioned above. It might also deal with emerging problems related to rapid technological change, such as the disposal of nuclear waste. Furthermore, if energy supplies were to be increased in the future, there is a need to make increased investments now. An oil fund which would operate along the lines of the International Monetary Fund, including conditionality, might be desirable. In sum, some kind of agreement between producers and consumers to create a more effective multilateral mechanism would be desirable to avoid further shocks.

Other participants felt that all of these envisaged arrangements would in reality be too difficult to negotiate and bring about; in other words, the practical difficulties should not be underestimated. In addition, there was some doubt that there would indeed be any shortages or price pressures at the end of the decade. There was extended discussion of this point, and although various reasons for a shortage were offered (increased demand, ending of the recession, etc.) there was certainly no consensus on this point. Some participants also felt that technological developments such as fusion power or solar power might solve the problem by itself, although in the very long run. There was finally no definitive answer to what, if anything, a multilateral agency concerned with energy might accomplish.

There was also no agreement as to the likelihood of warfare developing over energy sources, particularly oil. Some participants felt that there were distinct possibilities. Others felt that the sources themselves were not likely to be the cause of a major conflict. But conflicts over resources such as oil could easily become entangled with wider political conflicts to create an explosive international situation. However, it was more or less accepted that there seemed to be no immediate danger of a nuclear confrontation between the superpowers over the control of or access to energy sources.

The final issue addressed by the conference was food. The Chair (Joseph Black) pointed out that, in terms of multilateral cooperation, food has had a quite different history than the other two issues. The world has over 40 years of experience coping with food security problems multilaterally, and with the impact of hunger and malnutrition caused by distributional problems both between and within states.

One participant felt that food presented the best case for multilateralism. In this instance the national interest was not in conflict with the global interest. America's interests were in fact better served by multilateral solutions than by unilateral actions. But one of the main problems faced by American agricultural producers was the volatility of the world market. International reserves, therefore, could be used for the purposes of stabilization of both price and supply. Potentially, the United States budget deficit could be reduced if skyrocketing agricultural support costs could be reduced or restrained, and if surpluses could be disposed of in a cost effective manner. The fact that the United States has resisted multilateral approaches can be attributed to a free market ideology and to the belief that an essentially technical solution to increased agricultural production was feasible.

But many participants were of the opinion that the technical dimension was only one aspect of the food problem. Even though modern technology had the potential to greatly alleviate the food problem, it is doubtful if the technology can be effectively distributed and utilized in the most underdeveloped states, where it is most urgently needed. There is also, of course, a strong political dimension. The problems of Third World rural development, which are so intimately related to agriculture, ultimately require political solutions. Farmers and others living in rural areas need to acquire political power in order to reverse the urban bias in public policy which has been so evident in many developing countries. It was felt that a political approach would not only serve American interests; it would likewise promote economic stability in many parts of the Third World.

Other participants did not find this line of analysis entirely convincing. It was asserted, for example, that the United States need not accomplish the goal of price and supply stabilization through multilateralism (the intervention of an international organization into the system). Even though the United States could, if it wished, establish its own food reserve system for international purposes, at present it had no desire to solve the world food problem, which was a necessary first step toward a multilateral solution. A different conceptualization of the problem was therefore needed.

Still other participants felt that a middle approach was possible, i.e., that the United States could adhere to a multilateral solution without necessarily wanting explicitly to solve the world food problem. It was also contended that the two categories were not mutually exclusive. Any state could use both multilateral and other (unilateral and bilateral) approaches to further its national interests, and in fact the United States, particularly through the activities of non-governmental humanitarian agencies, has done so.

For many participants, the question of equity was paramount in the discussion of food. The phenomenon of severe malnutrition and even starvation in various parts of the world raises moral issues which can not be discussed solely within the context of any national interest.

The effect of food aid on agricultural production was another topic discussed with no definitive conclusion. Some studies have pointed out that certain developing countries have become dependent on food aid and that such aid allows governments to delay difficult national priority decisions. It was cautioned, however, that it would be unwise to generalize that it was thus undesirable for many countries to receive food aid.

In conclusion, the discussions concerning all three issues revealed that there is no obvious or simple answer to the question of how multilateral approaches to the management of resources can serve or even advance the interests of the United States. The discussions did, however, convey an acute awareness of the limitations of multilateral approaches. It was pointed out by the participants, for example, that such terms as "spaceship earth," "global planning" and "world order" were not used in this conference even though they were much in vogue only a few years ago. This is perhaps a reflection of the disillusionment in the United States with multilateralism in general, and with the United Nations and other international organizations in particular. An important question then is: is this trend permanent or temporary and, more importantly, is it justified? The chapters which follow hopefully will help to answer this question.

# PART I

## THE UNITED STATES AND THE MULTILATERAL MANAGEMENT OF ENERGY RESOURCES

# 1

# OIL AND CARTEL POWER

*Joan Edelman Spero*

At a time when it seemed that the South was becoming less important to the North and that frustration and anger were to be the lot of the underdeveloped countries, a ray of hope suddenly emerged. In 1973 Southern oil-exporting states dramatically reversed a system of dependency and increased not only their economic rewards but also their political power. The OPEC (Organization of Petroleum Exporting Countries) strategy, that of a producer cartel, became the new hope of the developing countries. (It was a short-lived hope when global recession, combined with overproduction, weakened the cartel's capacity to convert economic power into political power.)

## THE DEPENDENCY SYSTEM OF INTERNATIONAL OIL

The international oil system which was overturned by the producers was from its earliest days dominated by a small number of international oil companies.[1] These were the "seven sisters"--five U.S. (Standard Oil of New Jersey, now known as Exxon, Standard Oil of California, Mobil, Gulf, and Texaco), one British (British Petroleum), and one Anglo-Dutch (Royal Dutch-Shell).

This international oil oligopoly began before World War I in the United States with the Standard Oil trust and then spread throughout the world. A few companies, following the Standard Oil model, gained control of the domestic oil industry by the development of vertical

---

This chapter has been extracted from Joan Edelman Spero, The Politics of International Economic Relations, 2nd ed. (New York: St. Martin's Press, 1981), pp. 246-259. Reprinted with permission of the publisher.

integration, in this case, that is, by controlling all the supply, transportation, refining, and marketing operations. Vertical integration was reinforced by control of technology for exploration and refining.

In the late nineteenth century the large oil companies began to move abroad, and with their vast resources they obtained control of foreign supplies on extremely favorable terms.[2] However, their comfortable position was, in the years after World War I, threatened by ruinous competition. So, to preserve their dominance, the international oil companies adopted the familiar oligopolistic strategy of cooperation: to control supply, the seven sometimes shared sources through joint ventures and sometimes divided up sources by explicit agreements; to control their markets, they agreed to divide markets, fix world prices, and discriminate against outsiders.[3]

The home governments of the seven sisters played a supporting role in these efforts. The Northern imperial powers' political dominance of oil-producing regions facilitated oil company activities. Indonesia, for example, was controlled by the Dutch, the Middle East and North Africa by the British and the French, and Venezuela, somewhat less directly, by the United States. Aside from providing a favorable political and military environment, governments often took specific actions to support oil companies owned by their nationals. Under the San Remo agreement with the French in 1920, the British government attempted to restrict the oil rights in Mesopotamia to British and French companies. And, later, the U.S. government was of great assistance to U.S. oil companies in Venezuela and subsequently supported U.S. companies in Iraq and in the Persian Gulf at the expense of the British.

Faced with a few mammoth oil companies, which controlled technology and market access, cooperated with each other, and operated with the strong backing of their powerful home governments, developing countries were forced to bargain from a position of weakness. And, when the power of the companies was added to the uncertainty about the success of oil ventures and to the availability of alternative sources of supply, that bargaining relationship became even more distorted. As a result, in return for the payment of a small fixed royalty to host governments, the international oil companies obtained control over the production and sale of much of the world's oil.[4]

There was some evolution in this system in the decade following World War II. One important change, the impact of which was not felt until much later, was the emerging dependence of the developed market economies on imported oil. In the 1950s relatively inexpensive oil became the primary source of energy for the developed world. Western Europe and Japan, with no oil supplies of their own, soon became huge importers of oil; in the United States, oil consumption soon outdistanced even vast domestic production so that, by the end of 1950, the United States was a net importer of oil.

More evident in the early postwar era was a change in the attitude of host governments. Growing nationalism and awareness of the development gap engendered a more aggressive attitude toward the seven sisters, and decreasing uncertainty over the success of oil ventures and states' dissatisfaction with the terms of agreements negotiated under a very asymmetrical bargaining relationship engendered even more aggressive policies. In these years host governments succeeded in revising concession agreements negotiated before the war; first was Venezuela during the war, then Saudi Arabia in 1950, and, finally, the other Middle Eastern producing states renegotiated agreements with the companies. They redefined the basis for royalty payments and instituted an income tax on foreign oil operations. They also established what at the time was considered to be a revolutionary principle: that new royalties and taxes combined would yield a fifty-fifty division of profits between the companies and their respective host governments.[5] As a result, the profits accruing to host governments increased significantly. The per barrel payment to Saudi Arabia, for example, went from $0.17 in 1946 to $0.80 in 1956-57.[6]

Despite these important changes in Northern dependence and the Southern share of profits, the seven sisters continued to control the international oil system: They controlled almost all the world's oil reserves outside the Communist states, they controlled production at the wellhead, they dominated refining and transportation, and they controlled marketing. By these means they were thus able to manage the price of oil.

This continuing dominance of the seven sisters was based in part on their ability to preserve their oligopoly power. The incursion of competitors in upstream operations, that is, in crude oil production, was blocked by the majors' control through concession agreements of many oil-rich areas and by the long lead times inherent in finding and developing oil in territory unclaimed by the majors. Outsiders were also deterred from competing with the seven sisters downstream, that is, in the refining, transportation, and marketing operations. Not having their own crude oil supplies, competitors had to purchase oil from the majors. But, because the majors tended to take their profits at the upstream level, thus charging a high price for crude oil as compared with final product, there were small profits for downstream operations, which discouraged the entrance of competitors.

The ability of the oil oligopoly to manage the price of oil to their advantage was facilitated by the nature of the demand for oil. The demand for petroleum is highly inelastic, at least in the short run. Because there are no readily available substitutes and because it is difficult to decrease consumption, an increase in the price of oil does not significantly decrease the demand for oil. Thus, if companies can maintain a higher price for oil, they will not necessarily lose in appreciable sales volumes and so will continue to reap high profits. By controlling the supply of oil, companies could thus assure a high price.

This is exactly what the seven sisters succeeded in doing. They retained control of supply by keeping out competitors, and they cooperated among themselves to restrict that supply by a series of cooperative ventures: joint production and refining arrangements, long-term purchase and supply agreements, joint ownership of pipelines, and some joint marketing outside the United States. They also refrained from price competition.

Interestingly, maintaining prices at a level above that which would have prevailed under competitive conditions met with little resistance from the developed countries. European opposition to lowering the price of oil was squelched by a desire to protect the domestic coal industry, as lower oil prices would have increased oil consumption and thus have threatened the coal industry and coal miners, both important political groups. U.S. opposition was balanced by a concern for the domestic oil industry as well as by the political power of the five U.S. sisters.[7] U.S. oil was much more costly than foreign oil, and the domestic producers needed protection from lower international prices to survive. Thus the interests of the domestic and the international U.S. oil companies, both powerful political forces in the United States, supported a price which was high in terms of the cost of production.[8]

Finally, the dominance of the seven sisters was backed by political intervention, from the industry as well as from their home governments. One extreme example of such intervention took place in Iran in the early 1950s. In the period of renegotiation of concession agreements after World War II, the government of Iran sought a new agreement with the Anglo-Iranian Oil Company (AIOC, a predecessor of British Petroleum). When company-government talks broke down in 1951, the Iranian government of Premier Muhammed Mossadegh nationalized the AIOC assets in Iran. The British government became actively involved as an advocate of the company, of which it was a part owner. The British represented the company in negotiations and went so far as to impose a financial and economic embargo on Iran and to threaten military intervention. The British embargo, combined with an embargo on Iranian oil by the international majors, placed severe economic and political constraints on Mossadegh's government, accentuating the already existing internal political conflict over the role of the Shah. The United States at first sought to mediate between Britain and Iran. But, when numerous efforts failed and when Iran rejected a final offer which the United States considered reasonable and fair, the United States turned against Mossadegh. In August 1953 the opponents of Mossadegh, including the Shah with assistance from the CIA, overthrew Premier Mossadegh. A new concession was soon negotiated, and the political pressure from the oil companies and the consumer governments ended.[9]

## FORCES OF CHANGE

Several important changes, however, eventually altered the power of the international oil companies and led first producing and then consuming states to alter their dependence on the seven sisters.[10]

Changes occurred initially in the oligopolistic structure of the international oil industry. Competition increased in refining and marketing as an increasing number of refineries were built by private and public companies. At the same time competition for concessions to explore for and produce crude oil increased. Beginning in 1954, when smaller U.S. firms with the support of the U.S. government joined the new Iranian consortium, new companies became active on the international oil scene. Companies which had little or no international business--private U.S., European, and Japanese firms and French and Italian state-owned companies--began to seek concessions in existing and new oil-producing regions. They went to Algeria, Libya, and Nigeria, and they were successful. In 1952 the seven majors produced 90 percent of crude oil outside North America and the Communist countries, but by 1968 they produced only 75 percent.[11]

Changes also took place in the company control of pricing. Increased competition for supplies meant increased production. By the end of the 1950s, the growth in production outdistanced the growth in consumption. The seven sisters, no longer able to restrict supply, were unable to maintain the price of oil at the old level.

Aggravating the increasing supply were U.S. quotas on the import of foreign oil. These quotas were instituted in 1958 ostensibly to protect national security. Protection of the U.S. market from lower-priced foreign oil was necessary, it was argued, to ensure domestic production and national self-sufficiency.[12] In fact, such protection also conveniently served the economic interests of the domestic U.S. producers which could not have survived without protection, as well as the interests of the five U.S. internationals that also owned domestic sources of supply.[13] The effect of the quotas was to cut off the U.S. market for the absorption of the new supplies produced abroad. As a result, in 1959 and 1960 the international oil companies lowered the posted price of oil, the official price on which taxes are calculated. This act was to be a key catalyst for producer government action against the oil companies.

While changes in the structure of the international oil industry were undermining the dominance of the seven sisters, changes in oil-producing states were improving their bargaining position vis-à-vis the companies. The same factors which generally improved the position of host countries vis-à-vis foreign investors operated in the oil-producing states: changing elite attitudes, improved skills, decreasing uncertainty, and increasing alternatives.

The importance of the increase in alternatives was evidenced in the concessions negotiated by the new international oil companies. The emergence of these new participants increased the alternatives available to producer states and therefore increased the bargaining power of host governments. Producer states were thus able to obtain larger percentages of earnings and provisions for relinquishment of unexploited parts of concessions.[14] As a result of these negotiations, oil-producing governments increased their earnings and began to accumulate significant foreign exchange reserves. For example, from 1964 to 1970 the total official reserves of the major oil-exporting countries (Algeria, Indonesia, Iran, Iraq, Kuwait, Libya, Nigeria, Saudi Arabia, Trinidad and Tobago, and Venezuela) increased from $2.87 billion to $5.17 billion, and in 1971 the figure jumped to $8.18 billion. Furthermore, these reserves were concentrated in a few countries, particularly Libya and Saudi Arabia.[15] Monetary reserves meant that the important oil producers could afford the short-term loss of earnings that might result from an embargo or production reduction carried out in an effort to increase the price of oil or to seek other concessions from the companies and the consuming states.

Cooperation also improved the bargaining position of the host governments. What precipitated this cooperation were the price cuts of 1959 and 1960. Because their tax receipts are determined by price, the oil-exporting countries were greatly upset by these unilateral actions. So, in September 1960, five of the major petroleum-exporting countries--Iran, Iraq, Kuwait, Saudi Arabia, and Venezuela--met in Baghdad to discuss the oil company action. At that meeting the five decided to form OPEC to protect the price of oil and the revenues of their governments.[16] In its first decade OPEC expanded its membership from five to thirteen members who accounted for 85 percent of world oil export.[17] Despite the dominance of OPEC members in the world oil market, the new organization had little success in bargaining with the oil companies. The ability of OPEC to influence price or government revenues depended on the ability of the various members to cooperate to reduce production and thus to force a price increase. Although OPEC tried, it was unable to agree on production reduction schemes. It was not until the 1970s, when other conditions became favorable, that OPEC would emerge as a powerful and effective tool of the producer states.

A final and critical change in the international political economy of oil was the altered position of the Western consuming countries. As their reliance on oil as a primary source of energy grew and as internal sources of supply in the United States declined, the developed market economies became increasingly dependent on foreign oil, especially on oil from the Middle East and North Africa (see Table 1-1). By 1972 Western Europe derived almost 60 percent of its energy from oil, almost all of which was imported. Oil supplied 73 percent of Japan's energy needs, and virtually all of that oil came from abroad. Even the United States was becoming vulnerable. By 1972 the United States used oil for 46 percent of its energy and imported almost one-third of that required oil.

Table 1-1. Energy Consumption and Oil Consumption and Imports in the United States, Western Europe, and Japan, 1962 and 1972

| | 1962 | | | 1972 | | |
|---|---|---|---|---|---|---|
| | United States | Western Europe | Japan | United States | Western Europe | Japan |
| | Totals (millions of barrels per day) | | | | | |
| Energy consumption (oil equivalent) | 23.27 | 13.96 | 2.25 | 35.05 | 23.84 | 6.58 |
| Oil Consumption | 10.23 | 5.24 | 0.96 | 15.98 | 14.20 | 4.80 |
| Oil imports[a] | 2.12 | 5.19 | 0.98 | 4.74 | 14.06 | 4.78 |
| Middle East/ North Africa[b] | 0.34 | 3.80 | 0.72 | 0.70 | 11.30 | 3.78 |
| Other | 1.78 | 1.39 | 0.26 | 4.04 | 2.76 | 1.00 |
| | As % of Total Energy Consumption | | | | | |
| Oil Consumption | 44.0% | 37.5% | 42.7% | 45.6% | 59.6% | 73.0% |
| Oil imports[a] | 9.1 | 37.2 | 43.6 | 13.5 | 59.0 | 72.6 |
| Middle East/ North Africa[b] | 1.5 | 27.2 | 32.0 | 2.0 | 47.4 | 57.4 |
| Other | 7.6 | 10.0 | 11.6 | 11.5 | 11.6 | 15.2 |
| | As % of Total Oil Consumption | | | | | |
| Oil imports[a] | 20.7 | 99.0% | 102.1% | 29.7% | 99.0% | 99.6% |
| Middle East/ North Africa[b] | 3.3 | 72.5 | 75.0 | 4.4 | 79.5 | 78.6 |
| Other | 17.4 | 26.5 | 27.1 | 25.3 | 19.4 | 20.9 |
| | As % of Total Oil Imports | | | | | |
| Oil imports[a] | | | | | | |
| Middle East/ North Africa[b] | 16.0% | 73.2% | 73.5% | 14.9% | 80.4% | 78.9% |
| Other | 84.0 | 26.8 | 26.5 | 85.1 | 19.6 | 21.1 |

[a]Imports are gross rather than net; that is, exports are not deducted. Thus, they exclude product exports from West European refineries. For Japan, excess of imports over consumption arises because of small quantities of product exports, refinery losses, and (presumably) independent construction of the two series. By showing gross rather than net imports, the degree of foreign dependence is slightly overstated. The overstatement matters, if at all, only in the case of Western Europe.

[b]Includes negligible quantities from West Africa in 1961.

SOURCE: Joel Darmstadter and Hans H. Landsberg, "The Economic Background," in "The Oil Crisis: In Perspective," Daedalus, 104 (Fall 1975):21.

For Western Europe and Japan, the Middle East and North Africa became the overwhelmingly important source of oil imports. In 1972, 80.4 percent of Western European oil imports and 78.9 percent of Japan's imports came from this region. By 1972 even the United States relied on the

Middle East and North Africa for 14.9 percent of its oil imports. The developed market economies had become increasingly vulnerable to the threat of supply interruption or reduction. And this vulnerability was accentuated by the declining political influence of these countries' governments in the oil-producing regions and by the absence of individual or joint energy policies to counter any manipulation of supply.

As can be seen, in the 1960s and early 1970s major modifications gradually developed in the bargaining position of the three major actors in the international oil system. The position of the oil companies and of the developed market economies was significantly weakened, whereas the position of the producing states, especially those in the Middle East and North Africa, was significantly strengthened.

## The Process of Change: From Negotiation to Unilateral Power

Only minor changes occurred in the power structure of international oil in the 1960s. But, through a series of concessions, the oil-producing states succeeded in increasing their governments' revenues. The posted price of oil was never again lowered, but then it was never raised in this period to the pre-1959 level. And the oil-producing states gradually developed their ability and their experience in cooperation.[18]

The major changes in the system began in 1970. In the 1970s the oil-producing states--especially the Arab oil producers--developed a strong bargaining position vis-à-vis both the companies and the consuming states. Favorable international economic and political conditions plus internal cooperation enabled them, first, to take control of prices and, then, to assume actual ownership of oil investment.

The catalyst for future change was the Libyan government's action against the oil companies in 1970.[19] With Libyan oil supplying 25 percent of Western Europe's total oil imports, Libya had significant advantages in its bargaining relationship with the corporations.[20] Libyan oil was also important to certain independent oil companies which relied heavily on Libyan oil for their European markets and whose vulnerability separated them from the majors in Libya and thus prevented a common company bargaining position. Libya's bargaining position was further strengthened by its significant official foreign exchange reserves.

After the radical government of Colonel Muammar al-Qaddafi came to power in a coup in September 1969, it demanded negotiations with the oil companies for an increase in the posted price of and the tax on Libyan oil. Furthermore, when the talks stalled, the Libyan government threatened to nationalize the companies and cut Libyan oil production. As its first target the government selected the vulnerable Occidental Petroleum, which relied totally on Libya to supply its European markets and would have lost these markets without Libyan oil. Production reductions were imposed on Occidental in June and August 1970. In September Occidental capitulated, and the other companies were forced to follow.

The Libyan settlement was a harbinger of things to come. It provided for an immediate increase of $0.30 in the posted price of oil, an increase to $2.53 for Libyan crude, and for an increase of $0.02 a year until 1975 when the posted price would be $2.63. There was, in addition, an increase of approximately 5 percent in the income tax on oil. The Libyan contest with the oil companies had an important demonstration effect on the attitudes of other oil-producing states. It revealed the vulnerability of the companies and the unwillingness of the Western consumers to take forceful action in support of the companies. It also led to a major conceptual change in the minds of oil producers by demonstrating that government revenues could be raised not only by increasing exports but also or instead by increasing price.

In December 1970 the OPEC countries met in Caracas and passed resolutions calling for an increase in the posted price of oil and in host country income taxes on that oil. The companies in return sought to stem the tide. Their hope was to avoid a producer policy of divide and conquer and to enter into negotiations with all oil-producing countries to obtain a long-term, stable agreement on price and tax increases.[21] The governments of the oil-consuming states--although increasingly concerned about the trends in host-company relations--continued to allow the companies to manage relations with the oil producers.[22]

Negotiations between OPEC and the oil companies began in Teheran, Iran, in January 1971. When negotiations bogged down in February, the countries threatened to enact changes unilaterally and to cut off oil to the companies. The threat forced the companies to settle. The agreement at Teheran provided for an increase in the posted price of Persian Gulf oil from $1.80 to $2.29 per barrel (see Figure 1-1). To account for expected inflation, the agreement provided for an increase of 2.5 percent annually in this price. In addition, government royalties and taxes rose from 50 percent to 55 percent. In return, the companies received a five-year commitment on price and government revenues. In April 1971 a similar agreement, but with a higher price, was reached with Libya at Tripoli.

Barely four months later, when President Nixon announced the inconvertibility of the dollar and the 10 percent surcharge, a fundamental pillar on which the five-year agreements were based collapsed. Because the price of oil is stated in dollars, the real value of the Teheran and Tripoli agreements was altered when the value of the dollar was changed. After the Smithsonian agreement, the oil producers demanded a new agreement to account for international monetary changes. In January 1972 in Geneva an agreement was negotiated which provided for an increase in the posted price of oil and a continuing adjustment to

Figure 1.1. Changes in Posted Price of Crude Oil in Saudi Arabia[a], Iran[b], and Libya[c] 1970-1974

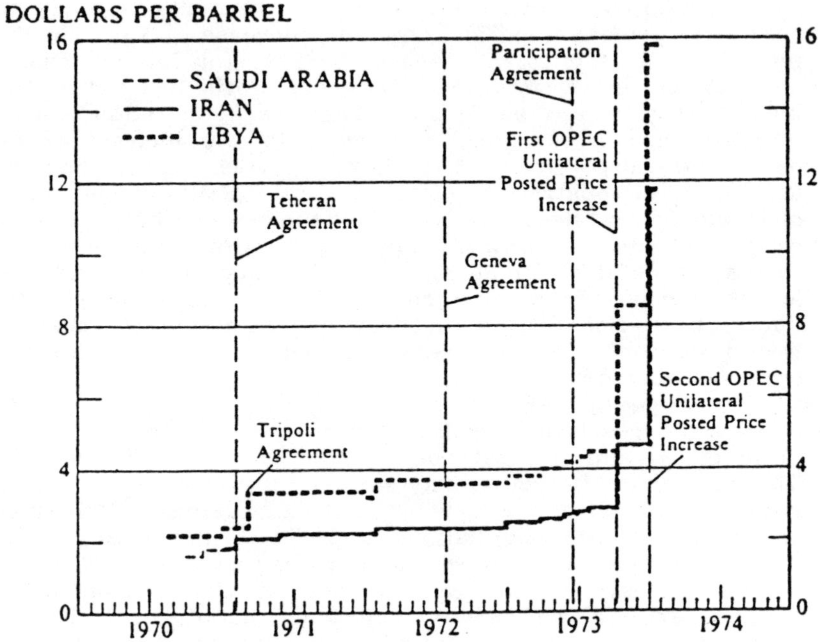


[a]Arabian Light 34°

[b]Iranian Heavy 27°

[c]Libyan Light 40°

SOURCE: Multinational Corporations and United States Foreign Policy, Hearings before the Subcommittee on Multinational Corporations of the Committee on Foreign Relations Part 5, U.S. Senate, 93rd Congress, 1st and 2nd Sessions, p. 268.

---

account for exchange rate changes. The price of Persian Gulf oil then rose to $2.48.

Despite this modification, the five-year period of stability sought by the companies--promised, they thought, at Teheran and Tripoli--proved ephemeral. No sooner had the issue of price and revenue been settled than the new issue of participation emerged. In 1972 OPEC called for a new conference to discuss nationalization. The companies resisted. OPEC, they argued, had no right to demand a reordering of the international oil system before 1975. But OPEC insisted that the question of participation was a totally new subject and therefore was not covered by Teheran and Tripoli.

Thus negotiations were held, and an agreement was reached in December 1972 among Saudi Arabia, Qatar, Abu

Dhabi, and the companies which provided a framework for the producing governments to negotiate with the oil companies in acquiring ownership of the companies operating in their states. Ownership would start at 25 percent and rise gradually to 51 percent of control by 1982.[23] Individual states then entered into negotiations with the oil concessionaires.

The final negotiation between oil producers and oil companies took place in October 1973. Despite significant achievements, the oil producers remained dissatisfied. The rising demand for oil in the consuming countries surpassed existing capacity[24] (see Figure 1-2). Resulting oil shortages drove up the market price of oil. However, because the posted price was fixed by the five-year agreements, it was the oil companies and not the oil producers that benefited from the rise in the price. Another reason for dissatisfaction was the fact that companies were making bids for new government-owned oil at prices above those of the Teheran and Tripoli agreements. Finally, increasing inflation in the West, the second devaluation, and the downward

Figure 1.2. World Crude Oil Spare Capacity and Demand (excluding Communist countries)

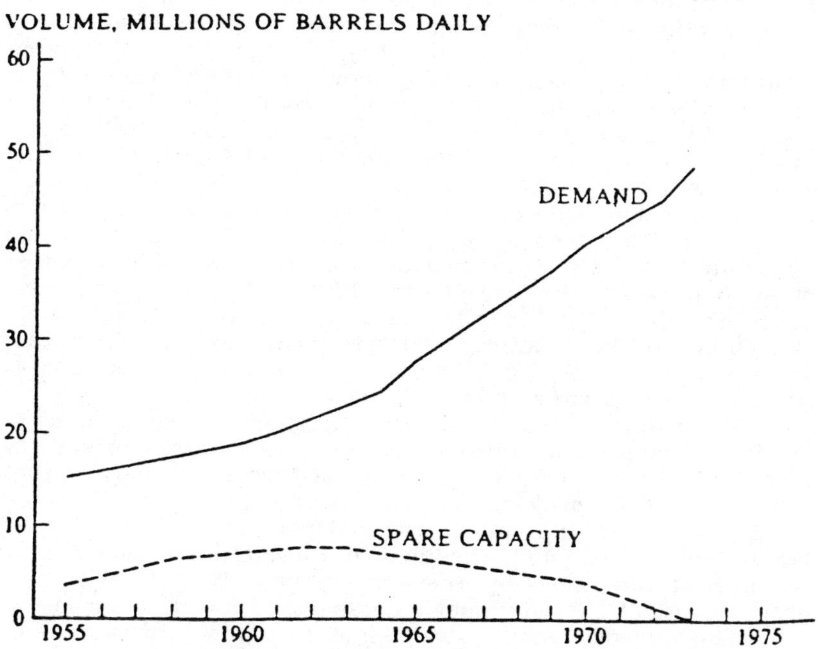


SOURCE: Multinational Corporations and United States Foreign Policy, Hearings before the Subcommittee on Multinational Corporations of the Committee on Foreign Relations, Part 5, U.S. Senate, 93rd Congress, 1st and 2nd Sessions, p. 212.

float of the dollar undermined the real value of earnings from oil production. Despite a second adjustment of prices to reflect monetary changes, the countries remained dissatisfied.

The conditions creating discontent also created the possibility for alleviating that discontent. Economic conditions were highly favorable to OPEC. Rapidly rising demand and shortages of supply meant that the developed market economies were highly vulnerable to supply interruption by even one significant producer and were therefore in a weak bargaining position on price. In September 1973, when OPEC summoned the oil companies for negotiations, they came. Negotiations began on October 8 in Vienna. The oil producers demanded substantial increases in the price of oil, the companies stalled, and on October 12 the companies requested a two-week adjournment of talks to consult with their home governments.

The adjournment, it turned out, was not for two weeks but forever. Political as well as economic conditions now enhanced the bargaining position and escalated the demands of the most powerful of the oil producers: the Arab states. The fourth Arab-Israeli war had begun on October 6, just two days before the Vienna talks began. A common interest in supporting the Arab cause vis-à-vis Israel and supporters of Israel in the consuming states was an important force for unity of the Arab members of OPEC in their economic confrontation with the companies and the consumers. On October 16 the Organization of Arab Petroleum Exporting Countries (OAPEC) unilaterally increased the price of their crude oil--to $5.12.[25] Other oil producers followed with comparable price increases. On December 23, OPEC unilaterally raised the price of Persian Gulf oil to $11.65 to take effect January 1, 1974.

The actions of October and January demonstrated that, henceforth, on issues of price the producers need not negotiate with the companies or with the consuming states; they possessed the power to dictate terms unilaterally. Since the fall of 1973, oil prices have been determined by the OPEC members acting as a producer cartel. Producers have negotiated among themselves to determine the posted price of oil and have then imposed that price on consumers. OPEC members have instituted production reductions to manipulate supply conditions and thus to maintain price, the same technique used by the seven sisters in an earlier era. Crucial to this program of maintaining price by reducing supply has been the role of the major reserve countries and large producers: Saudi Arabia and Kuwait have been willing to support the cartel by themselves, absorbing a large part of the production reductions necessary to maintain price.[26]

The oil companies [were unable] to resist the transfer of power from company to host government. The monopoly control of oil by OPEC, the unity of the producers, and tight market conditions . . . prevented the companies from having any leverage. Furthermore, the companies [had] little incentive to resist. They [had] been able in most cases to pass price increases along to consumers and thus

[had] not suffered financially from the loss of control over price.

Producers . . . asserted their power not only on questions of price but also on issues of ownership. The participation agreement proved to be as fleeting as those in Teheran and Tripoli. The new sense of power over price was quickly translated into a desire for immediate and total equity control. All the major oil-producing states have now signed agreements with the oil companies providing for immediate majority or total national ownership of subsidiaries located in those states.

Thus the international oil companies have lost their powerful position as managers of the international political economy of oil. It would be wrong, however, to think that they have disappeared. Although no longer the arbiters of supply and price, although no longer the owners of oil concessions, the seven sisters and their many smaller relatives still play a vital role in the international oil scene. As holders of technology and markets, they are needed by the newly powerful producer governments. As their holdings are nationalized, their role is becoming that of a vitally necessary service contractor to producer states. Still, it is a far cry from the days when the companies divided up the producing regions among themselves and obtained control of the world's oil for a song.

With the decline of the companies, Northern consumer governments attempted to counter the power of the producers.[27] But the consumers, unlike the producers, were unable to agree on a common policy. The United States sought to form a countercartel to destroy OPEC. U.S. policy makers reasoned that unified opposition to OPEC and the threat of economic or military retaliation would undermine producer solidarity and lead to a decline in oil prices. To this end, the United States pressed Western Europe and Japan to attend a consumer conference and to develop a united policy. The conference, held in Washington in February 1974, demonstrated the inability of the consumers to agree on a common strategy. While the United States argued for unity and resistance, the Europeans and Japanese--far more dependent on foreign sources of oil, far less interested in support for Israel, and somewhat fearful of U.S. dominance--opposed a countercartel and argued for cooperation with the producers. The only common effort to emerge from the conference was an agreement to establish an International Energy Agency (IEA) to work on an emergency oil-sharing scheme and a long-term program for the development of alternate forms of energy.

After the Washington conference, the United States, the Europeans, and the Japanese went their own ways. The United States sought to destroy producer unity and to divide the rest of the Third World from OPEC by continuing to press for consumer unity and the development of the IEA. The Europeans sought special bilateral political and economic arrangements with the oil producers and resisted consumer bloc strategies. France, the strongest opponent of the U.S. approach, refused to join the IEA and urged instead a producer-consumer dialogue.

In late 1974 a compromise was reached between the United States and France. The United States obtained grudging French acceptance of the IEA, although France refused to join, and France obtained grudging U.S. support for a producer-consumer dialogue. After some delays and conflicts, the Conference on International Economic Cooperation (CIEC) finally commenced in December 1975. The conference constituted a recognition, if not a total acceptance, by the United States and the other consumers that they could not alter the power of the producer states over the price of oil and that what they could seek at best was some conciliation and coordination of common interests. Even that coordination proved elusive. In 1977 CIEC and the effort to achieve a forum for a producer-consumer dialogue failed.

## NOTES

1. The following analysis of the international oil oligopoly is based largely on accounts in Morris A. Adelman, *The World Petroleum Market* (Baltimore, Md.: Johns Hopkins Press, 1972); J. E. Hartshorn, *Politics and World Oil Economics: An Account of the International Oil Industry in Its Political Environment*, rev. ed. (New York: Praeger, 1962 and 1967); Edith T. Penrose, *The Large International Firm In Developing Countries: The International Petroleum Industry* (Cambridge, Mass.: M.I.T. Press, 1969); Anthony Sampson, *The Seven Sisters: The Great Oil Companies And The World They Made* (New York: Viking Books, 1975); Federal Trade Commission, *International Petroleum Cartel*, Staff Report to the Federal Trade Commission submitted to the Subcommittee on Monopoly of the Select Committee on Small Business, U.S. Senate, 82d Congress, 2d session (Committee Print No.6)(Washington, D.C., 1952); and John M. Blair, *The Control Of Oil* (New York: Pantheon Books, 1976).

2. See Zuhayr Mikdashi, *A Financial Analysis Of Middle Eastern Oil Concessions: 1901-1965* (New York: Praeger, 1966); Charles Issawi and Mohammed Yeganeh, *The Economics Of Middle Eastern Oil* (New York: Praeger, 1962), pp. 24-40.

3. In 1928, for example, Shell, Standard Oil, and Anglo-Persian (the predecessor of BP) concluded the "As Is," or "Achnacarry," agreement to divide world markets and determine world prices, and in that same year a group of British, Dutch, U.S., and French companies agreed to divide up much of the old Ottoman Empire in the Red Line agreement. Also important was the basing-point pricing system, which established a common price at several locations, or basing points, and standard, not actual, freight charges from the basing point to the destination. This

system prevented low-cost producers from expanding their market share by reducing prices. See Penrose, The Large International Firm In Developing Countries, pp. 180-83.

4. See Mikdashi, A Financial Analysis; Issawi and Yeganeh, The Economics Of Middle Eastern Oil; Gertrude G. Edwards, "Foreign Petroleum Companies and the State in Venezuela," in Raymond F. Mikesell et al., Foreign Investment In The Petroleum And Mineral Industries (Baltimore, Md.: Johns Hopkins Press, 1971), pp. 101-28; Franklin Tugwell, The Politics Of Oil In Venezuela (Stanford, Calif.: Stanford University Press, 1975); Donald A. Wells, "Aramco: The Evolution of an Oil Concession," in Mikesell et al., Foreign Investment, pp. 216-36.

5. Wells, "Aramco: The Evolution of an Oil Concession."

6. Adelman, The World Petroleum Market, p. 207.

7. See, for example, Robert Engler, The Politics of Oil: Private Power and Democratic Directions (Chicago: University of Chicago Press, 1961).

8. This interest later led to support for an oil import quota system imposed in 1958 and designed to protect the U.S. market from cheap foreign oil.

9. Benjamin Shwadran, The Middle East, Oil and the Great Powers (New York: Council for Middle Eastern Affairs, 1955), pp. 103-52; J. C. Hurewitz, Middle East Politics: The Military Dimension (New York: Praeger, 1969), pp. 281-82.

10. See Penrose, The Large International Firm in Developing Countries, pp. 248-63; Adelman, The World Petroleum Market, pp. 196-204.

11. Mira Wilkins, The Maturing of Multinational Enterprise: American Business Abroad from 1914 to 1970 (Cambridge, Mass.: Harvard University Press, 1974), pp. 386-87.

12. See, for example, Joan Edelman Spero, "Energy Self-Sufficiency and National Security," in The National Energy Problem, eds. Robert H. Connery and Robert S. Gilmour Proceedings of the Academy of Political Science, 31 (December 1973):123-36.

13. See, for example, Engler, The Politics of Oil.

14. See footnote 4.

15. International Monetary Fund, Annual Report of the Executive Directors for the Fiscal Year Ended April 30,

1972 (Washington, D.C.: International Monetary Fund, 1972), pp. 31-32.

16. For a history of the Organization of Petroleum Exporting Countries, see Zuhayr Mikdashi, The Community of Oil Exporting Countries: A Study in Governmental Cooperation (Ithaca, N.Y.: Cornell University Press, 1972).

17. Zuhayr Mikdashi, "The OPEC Process," Daedalus, 104 (Fall 1975):204. The new members were Abu Dhabi, Algeria, Libya, Qatar, the United Arab Emirates, Nigeria, Ecuador, Indonesia, and Gabon.

18. Mikdashi, The Community of Oil Exporting Countries, pp. 196-207.

19. On evolution of events in Libya, see U.S. Senate Committee on Foreign Relations, Multinational Corporations and United States Foreign Policy: Multinational Petroleum Companies and Foreign Policy, hearings before the Subcommittee on Multinational Corporations, 93rd Congress, 1st and 2nd sessions, pt. 5 (1974).

20. Organization for Economic Cooperation and Development, Oil Committee, Oil Statistics: Supply and Disposal 1970 (Paris: Organization for Economic Cooperation and Development, 1971), p. 27. There were several reasons for the powerful Libyan position in the European market. Transportation of oil from Libya was much cheaper and safer than transportation of oil from the Persian Gulf, which, with the closing of the Suez Canal, required a long trip around Africa. Furthermore, in 1970 there had been a decline in the supply of oil from Nigeria due to the civil war, and the pipeline which carried Saudi oil to the Mediterranean had been cut in Syria. Finally, Libyan oil was low in sulfur and therefore desirable for environmental reasons.

21. See U.S. Senate, Multinational Corporations.

22. There was some consultation by the developed market states. The U.S. Department of Justice issued a waiver to oil companies under antitrust law enabling them to cooperate in bargaining to resist unreasonable demands for higher prices. See U.S. Senate, Multinational Corporations, pt. 5, pp. 145-73. President Nixon then sent Undersecretary of State John N. Irwin to the Middle East to encourage governments to enter into joint negotiations with the companies. Secretary Irwin, however, capitulated to the demand of the Shah of Iran for separate negotiations.

23. Kuwait signed the agreement, but the national assembly did not ratify it.

24. For the reasons, see Joel Darmstadter and Hans H. Landsberg, "The Economic Background," Daedalus 104 (Fall 1975):15-37.

25. The Organization of Arab Petroleum Exporting Countries was formed by three Arab states, Kuwait, Libya, and Saudi Arabia, in 1968. It was expanded in 1970 to include Algeria, Abu Dhabi, Bahrain, Dubai, and Qatar.

26. See Dankwart A. Rustow and John F. Mugno, Opec: Success and Prospects (New York: New York University Press, 1977).

27. See Wilfrid L. Kohl, "The United States, Western Europe and the Energy Problem," Journal of International Affairs 30 (Spring/Summer 1976):81-96.

# 2

# NO FUEL FOR DEVELOPMENT: UNITED STATES POLICY TOWARD MANAGING INTERNATIONAL ENERGY RESOURCES

*Richard J. Kessler*

## INTRODUCTION

Energy has been a major issue in international relations since the oil embargo of 1973-74 when it became an agenda item at the industrial nations' annual economic summits, at several special sessions of the United Nations General Assembly, and at ad hoc global conferences and meetings of the nonaligned nations. These meetings have resulted in the formation of new energy programs and agencies. At issue in all this activity have been, first, how to assist nations to secure economically and otherwise their energy supplies and, second, how to manage energy price increases.

Despite all these efforts, there have been no dramatic solutions. This lack of fundamental change is in part because the industrial nations, especially the United States, did not want to "politicize" energy--the essential objective of the developing nations. Working at cross-purposes, no middle ground could be achieved. The initial U.S. response, as one development specialist observed, was a "scheme of counter-confrontation." The United States attempted to organize the industrial nations, forming an International Energy Agency (IEA), titularly a branch of the Organization for Economic Cooperation and Development (OECD) in Paris, excluding the developing countries from membership: "The U.S. reaction was to protect its own

---

A version of this chapter was included in a portion of a 1984 report by the Energy and Strategic Resources Program of the Center for Strategic and International Studies, Georgetown University.

narrowly defined interests first and the interests of other industrialized nations only second."[1] At the time of the 1973-74 oil crisis the United States had for two decades been fighting Third World demands for greater control over their resources.

The core of the debate has been the concept of permanent sovereignty over natural resources. First raised in the United Nations in 1952 at the Eighth Session of the Human Rights Commission, while discussing preparation of the Draft International Covenants on Human Rights, delegates argued that "the right of peoples to freely dispose of their natural resources had to be recognized as an essential element of economic independence."[2] The developed countries were concerned about states nationalizing foreign investment, while developing countries wanted recognition of that right.

The debate culminated in 1962 with UN General Assembly Resolution 1803 (XVIII, December 14) on Permanent Sovereignty Over Natural Wealth and Resources. The right of "peoples and nations to permanent sovereignty over their natural wealth and resources" was asserted, but "appropriate compensation" had to be provided.[3] Since then, numerous UN debates have attempted to define more specifically what "sovereignty" and "compensation" meant.

The decade between 1952 and 1962 saw newly independent states asserting that "political sovereignty" was the critical determinant for their national development. Nationalism was the form that this assertion took and was necessary to loosening ties with previous colonial rulers and to solidifying the new nation-state. The preoccupation with political sovereignty in the 1950s did not exclude the attachment of importance to economic development--only it included the assumption that political sovereignty was a necessary condition for economic independence, later to be called self-reliance.

In the early 1960s, the failure of new states to achieve rapid growth rates and levels of development commensurate with expectations at independence led to the perception that political independence was not synonymous with economic independence. Economic sovereignty supplanted political sovereignty as a goal not because economic independence was perceived as more important than political independence but because it was assumed that without economic independence there could be no political independence. In the 1950s and 1960s the assertion of sovereignty over natural resources had been important, but in the 1970s it was access to resources which became critical.[4]

The economic and political events of the early 1970s altered the approach of the developing countries. The oil embargo of 1973 and the 400 percent oil price rise made the 1970s the "Energy Decade." The developing countries applauded OPEC's action to take control over its natural resources even though most developing countries had to pay a high price in terms of their oil imports. But that event showed that developing countries could have world economic power. Analysts predicted that other commodity cartels would be formed.[5] But the prophecy was false.

And in the beginning of the 1980s, the developing countries were paying the price for the oil price rises in enormous debt and low growth. Even the oil-producing nations were suffering as the high oil prices of the 1970s encouraged the development of non-OPEC energy resources and slowed the growth of the developing and developed countries.

In 1975 a number of new initiatives had been taken to manage this financial problem: Additional credit or grants were provided to the developing countries. They included a special International Monetary Fund (IMF) Oil Facility and Oil Facility interest subsidy, an IMF Trust Fund, a Development Security Facility (an enhanced IMF Compensatory Financing Facility), establishing a Third Window at the World Bank for low-interest loans, and a UN Emergency Fund.[6] The United States, however, did not participate in any of these activities. These efforts were geared to assisting developing countries in their initial efforts to pay for oil imports or to compensate them for the effects of those costly imports.

Most important were the efforts by commercial banks in the West to recycle the tremendous increase in funds flowing into their coffers from OPEC nations. The current account balance of OPEC nations had increased sevenfold. By 1979 this flood of surplus funds had almost evaporated. Much of this money went to the West as OPEC direct investment and purchases. But a great deal was recycled to the developing countries, some of whose economies during this period grew as much as three times the rate of the industrial states, at the same time quadrupling their external indebtedness. By 1979 the developing countries owed $400 billion, with about two-thirds of that amount lent by commercial banks at commercial interest rates.[7]

The danger of the cumulative impact of this debt became apparent in the early 1980s when persistent economic recession in the West reduced demand for imports, including energy, affecting the ability of developing countries borrowers to service their debt to Western banks and to international lending institutions. President Reagan's Secretary of the Treasury, Donald Regan, suggested that a new international financial institution might be necessary to cope with this enormous debt. In reaction, the international commercial banks established in Washington an institute to monitor foreign debt and to advise them. Unlike any other issue, financial problems brought a spirit of cooperation to the world that had not been witnessed since the 1944 Bretton Woods Conference which had led to the establishment in 1945 of the IMF and the International Bank for Reconstruction and Development (IBRD or World Bank), in the aftermath of World War II.

The last three decades could be entitled "From Independence to Interdependence." During that time frame, first one group of nations and then another attempted to establish dominance in a world in which dominance was no longer possible. The framework for the debate over energy resource issues of the 1970s was confrontational: North vs. South, Haves vs. Have-Nots, OPEC vs. West. Each side thought it would prevail, with sufficient bludgeoning, over the other. In every meeting one group organized against another.

Even in meetings of international financial institutions where substantial agreement did occur, there was a struggle on the part of the Western nations, principally the United States, to avoid a dilution of their voting power, and, thereby, control over the direction of the institutions themselves. There was, however, substantial agreement over the need for international cooperation.

The reason for this unusual, albeit reluctant, spirit of cooperation is instructive. The international financial problems of the first oil crisis and of the second, still being played out in the 1980s, affected every nation. The expansion of Western commercial banking both before the 1973 oil crisis and afterwards, partly in response to it, made the economic security of the Western states a direct function of the economic well-being of the developing countries, including OPEC, and vice-versa. Even the Reagan Administration, which took office decrying multilateral aid, was forced to reverse its position when confronted with the potential collapse of the U.S. financial community as a result of lending to the developing countries.

Energy resource development has not been characterized by the same spirit of cooperation. This is in part because conflict developed between a host country's interest in controlling its resource development and the foreign corporations eager to exploit them. It is also due in part to the differing attitudes toward private enterprise, not only in the developing countries but also in the developed. The United States, for example, is the last large industrial nation without a national oil company. Finally, it is due to the nature of the international petroleum industry which, although it does cooperate in consortia, tends to be highly competitive and individualistic. Although the large "Seven Sisters" (British Petroleum, Exxon, Gulf, Royal-Dutch-Shell, Mobil, Texaco, and Standard Oil of California) dominate much of the market, there are a multitude of lesser stars surrounding the Pleiades. In addition, even the majors do not always agree. This diversity makes it difficult for U.S. policy makers to formulate a consensus policy on cooperation with other states on energy. An example of this is the series of international meetings arranged after 1973 to discuss energy problems. Several different tracks were pursued simultaneously. The general intent of U.S. policy was to establish its leadership of the bloc of industrial states to oppose OPEC, while undermining efforts by the developing countries to make energy part of a larger agenda of grievances shared against the West.

## NORTH-SOUTH NEGOTIATING POSTURES: AN INTERPRETATIVE HISTORY

Since 1973 the struggle in international fora has been over whether to make energy a central issue or part of a larger package of issues, conveniently called the New International Economic Order (NIEO). The Western industrialized, or Group B countries, led by the United States, has attempted to make energy a separate development policy problem--while the developing countries, or Group of 77, have argued that it should be considered part of a larger policy problem. The Group of 77 has been largely successful in making the debate focus on holding global negotiations for a New International Economic Order. But because the Group B countries dominate the international economic system, it has been impossible to implement NIEO. On this issue, the developing countries have won some battles but lost the war.

Numerous international meetings have been held since 1973 which have provided opportunities to "push" energy as an international issue. Among the most important have been:

(1) Fourth Non-Aligned Summit in Algiers, September 1973.
(2) Washington Energy Conference, February 1974.
(3) Sixth Special Session of the UN General Assembly, April 1974.
(4) Conference on Raw Materials in Dakar (Senegal), February 3-8, 1975.
(5) UN Industrial Development Organization (UNIDO) III Meeting in Lima, March 12-26, 1975.
(6) Seventh Special Session of the UN General Assembly, September 1-16, 1975.
(7) Conference on International Economic Cooperation (CIEC) in Paris, December 1975-77.
(8) UN Conference on Trade and Development (UNCTAD) IV in Nairobi, May 5-31, 1976.
(9) Fifth Non-Aligned Summit, Colombo, December 1978.
(10) Preparatory Meeting for UNCTAD V in Arusha (Tanzania), December 1978.
(11) UNCTAD V in Manila, May-June 1979.
(12) Tokyo Economic Summit, June 28, 1979.
(13) World Conference on Agrarian Reform and Rural Development, Rome, July 1979.
(14) Sixth Non-Aligned Summit, Havana, September 1979.
(15) UNIDO III Meeting in New Delhi, January 21 through February 9, 1980.
(16) Venice Economic Summit, June 1980.
(17) Eleventh special session of the UN General Assembly, August 1980.
(18) Ottawa Economic Summit, July 1981.
(19) UN Conference on New and Renewable Sources of Energy (UNCNRSE), August 1981.

(20) Cancun Summit, October 1981.
(21) Versailles Economic Summit, June 1982.
(22) Seventh Non-Aligned Summit, New Delhi, March 1983.
(23) Williamsburg Economic Summit, May 1983.
(24) London Economic Summit, June 1984.

From the perspective of the Group B and Group of 77 countries, these meetings were tests of will at which each attempted to force through its particular agenda. The tactic of the United States was to place concrete proposals on the table which could only be implemented through existing institutional frameworks. The tactic of the Group of 77 was to propose sweeping reforms, requiring new agencies. Within the Group B, there also were varying degrees of sympathy for the Group of 77 demands.

At the fourth summit of the nonaligned countries in Algiers in September 1973, the concept of NIEO was developed. It resulted in a "Declaration on the Establishment of a New International Economic Order" at the Sixth Special Session of the UN General Assembly in April 1974 and was followed in December at the twenty-ninth UN General Assembly by a "Charter of Economic Rights and Duties of States." Early on, the developing countries took the initiative and the West by surprise by articulating and gaining acceptance of a wide program affecting all elements of the international economy, including energy. The West was slow to respond.

On January 9, 1974, the United States called for a conference of major industrial oil consuming countries warning those states against establishing bilateral agreements with OPEC. A February meeting held in Washington led to the formation of the International Energy Agency (IEA) in Paris. All the industrial nations, except for France, joined the IEA, whose initial objective was to form a consumer cartel in order to undermine OPEC.

The U.S. goal, as stated by Secretary of State Henry Kissinger, was short-term and relatively self-serving. It was "to build a series of policies which would . . . protect us against short-term dangers such as embargoes and the destabilizing movements of asset[s] held by oil countries."[8]

The key to this policy was the IEA Agreement, signed on November 19, 1974, which established an emergency program for sharing oil in any future crisis and other programs to monitor energy supplies and develop new resources. The 1973-74 Arab oil embargo had been directed primarily at the United States. Secretary Kissinger's inspired policy was to limit the effectiveness of any future embargo by forcing all the industrial nations (except for France which refused to join the IEA) to share oil supplies. When during President Reagan's Administration, total U.S. oil imports declined, especially those from the Middle East, the U.S. government became less supportive of the IEA. Even before Reagan's inaugural, General Alexander Haig, testifying before Congress on the IEA in 1980, observed:

> It [IEA] was launched with great hopes and with at least a symbol of cooperation even from France. Subsequently, in the months and years that followed, of course, far less has been done than I felt was necessary in the way of solid American follow up, and solid American leadership. . . .[9]

At the same time, at the Washington Energy Conference of industrial nations, Algeria appealed to the United States to convene a UN Special Session to discuss raw materials and development. This was agreed to by February 15.

The Sixth Special Session of the United Nations General Assembly, convened in April 1974, resulted in a draft Declaration on the Establishment of a New International Economic Order (NIEO) and a Program of Action for reconstituting the world's economic system.

The cohesion in the Third World necessary to agree on the NIEO and later to have the proposed Charter of Economic Rights and Duties of States to be accepted overwhelmingly at the twenty-ninth UN General Assembly on December 12, 1974 (only Belgium, Denmark, West Germany, the United Kingdom, and the United States opposed it; ten other industrial nations abstained) signalled to the United States the depth of opposition to its position. That signal was ignored until the collapse of another series of multilateral meetings (CIEC) the next spring. Only then did Secretary of State Kissinger shift the U.S. position.

The developing countries followed up their earlier successes with a conference on raw materials in Dakar, Senegal, in February 1975. This conference discussed the formation of OPEC-like commodity cartels and intergovernmental cooperation and consultation. The issues of commodity cartels and/or price stabilizations became an agenda item at each successive international meeting. In Lima, Peru in March 1975, the second General Conference of the UN Industrial Development Organization (UNIDO) was convened and resulted in the "Lima Declaration and Plan of Action on Industrial Development and Cooperation." While the Lima Declaration was open to all countries, the program or plan of action of the meeting had been defined at the previous Algiers meeting, the Sixth Special Session and the twenty-ninth annual session of the UN General Assembly. The Lima Declaration recommended that UNIDO become a UN Specialized Agency and that an Industrial Development Fund be established. Only the United States voted against the resolution--in essence taking the flak from the developing countries, while the other Group B countries voted in favor, secure in the knowledge that U.S. opposition would be sufficient to prevent progress.

Energy was an incidental topic at all these meetings although it was argued by the United States that energy was central to the economic issues at hand: commodity price stabilization, development, and aid and assistance.

Recognizing that debate in international fora was getting out of hand and that the United States was becoming isolated, the United States, led by Secretary of State Kissinger, prepared a counter-agenda, focusing specifically on energy.

In September 1975 the UN General Assembly's Seventh Special Session opened. Secretary Kissinger, in a dramatic speech entitled "Global Consensus and Economic Development," proposed a series of concrete measures which created a spirit of goodwill in the otherwise acrimonious international community. Most importantly he proposed an International Resources Bank (IRB) to deal with commodity issues and an international energy body. It was the first time that the United States had shown any interest in discussing commodity problems. The Secretary of State also called for the establishment of an International Energy Institute. The target of this proposal was both the UN Special Session of the General Assembly meeting and the forthcoming CIEC meeting. The United States wanted to reestablish its leadership before disarray among its allies forced them into greater cooperation with the developing countries and OPEC. By agreeing to talk, the United States delayed agreement (or disagreement) on substantive issues.

While the Seventh Special Session of the General Assembly was dominated by Secretary of State Kissinger's speech with its detailed proposals, the developing countries had also come well prepared. By adopting a conciliatory approach, the United States avoided confrontation at perhaps the Third World's greatest moment of strength. Discussions were held on specific issues, as would occur in CIEC, while avoiding agreement on a more comprehensive NIEO. The United States agreed to contribute to a UN Revolving Fund for Natural Resource Exploration. However, energy was still not the focus of discussions. In Secretary of State Kissinger's words:

> We decided in the early part of this year [1975] that we would make a maximum effort to conduct our relations, not only with the [oil] producers but with developing countries on the basis of understanding, of mutual respect, and very concrete proposals. We wanted to avoid ideological confrontation. . . .[10]

The proposed International Energy Institute never got off the ground. It was never clear what the Institute would do or by whom and by how much it would be funded. Some saw it as a clearinghouse for information on energy technologies, while others saw it more as an IEA-type organization. The U.S. Secretary of the Treasury, William Simon, opposed the Institute as being in competition with private sector initiatives.

The UN meeting was followed in October by a second preparatory session (the first was in April) of the Paris Conference on International Economic Cooperation. CIEC had originated in an October 1975 proposal by France's President Valery Giscard d'Estaing to hold an oil conference to which oil consumers and producers would be invited. Its objective was to replace confrontation with cooperation.

The CIEC discussion, which ran from December 16, 1975, to June 2, 1977, involved 27 countries. Eight came from the West and of the 19 from the developing countries, 7 were OPEC members. While agreement was reached on 20 issues, participants disagreed on 21. Four commissions were formed on Energy, Raw Materials, Development, and Finance with the participation of 8 developed countries (including an EEC delegation) and 19 developing countries. The Energy Commission addressed such questions as energy supply and demand, and international cooperation. At its conclusion only two substantive proposals had been agreed upon: a Special Action Program to assist the "least developed of the developing countries" with $1 billion of additional aid and a recommendation that the World Bank place priority on energy loans.[11]

CIEC was viewed by many countries as a delaying action on the part of the United States to prevent real action on issues. The United States viewed it as an effort to attain real progress by dividing issues into practical parts. But U.S. policy in other arenas was designed to prevent substantive agreement.

The U.S. negotiators were well aware that any agreement would require U.S. support to proceed but that the United States could not completely "stonewall" any negotiation without enraging its allies and isolating the United States diplomatically. As Secretary of State Kissinger stated: "Without our leadership and our example, there can and will not be an effective foreign energy policy."[12] The United States' European allies were, for the most part, more favorably disposed toward negotiations with OPEC and the other developing countries on energy and a broad range of issues. Even though Secretary of State Kissinger had called for a producer-consumer dialogue as early as December 12, 1973, in a speech in London,[13] the United States was not willing to discuss the main item on the developing countries agenda: a redistribution of resources from North to South.

The April 1975 preparatory meeting in Paris for CIEC was considered a failure because of U.S. intransigency. The United States delayed the talks by not compromising, expecting a recession induced by rapid increases in oil prices would drive down oil demand and break up OPEC. A U.S. delegate characterized the West's strategy as gaining control of the oil market in order to "hasten OPEC's demise."[14] Although these preliminary negotiations failed, another preparatory meeting was convened in October, resulting in an agreement to hold CIEC.

The change in the U.S. position came from fears of division within the West and the realization that OPEC would not dissolve, nor would crippling animosity develop in relations between OPEC and the Third World. In May 1975, Secretary of State Kissinger gave a speech in Kansas City stating that the United States would alter its "free market approach."[15] Testifying in September, he stated that the United States opposed:

> a tendency toward a comprehensive approach to all the economic problems of the developed and developing countries in one big forum, simultaneously. Our policy has always been that we favor a dialogue that gives specific objectives with concrete foreseeable outcomes and not a confrontation between blocs.[16]

CIEC was considered by some a success and others a failure. It depends upon what was perceived as its objective. One observer stated the "CIEC materialized precisely because the West bowed to OPEC demands that energy and North-South issues be linked."[17] CIEC added to OPEC's legitimacy, made the West more flexible toward developing countries demands, and stabilized relations between the North and the South, including OPEC.[18] However, as another forum for global negotiations it failed: "The pattern at the CIEC talks, the Seventh Special Session, and UNCTAD IV appeared as much talk, many pieces of paper, and virtually no change in the status quo."[19] If the primary U.S. objective was to avoid agreement on changing the international economic order, it had succeeded. But by agreeing to discussions, it provided a forum and thereby a degree of legitimacy to Third World demands.

Following the end of the sessions, the UN General Assembly created a Committee of the Whole (COW) in 1977 to secure follow-up action to CIEC. The United States hoped the COW would provide a high-level forum at which talks could be conducted at a ministerial level. Rather than focusing on global negotiations, the United States wanted discussions to focus on such issues as alternative energy resources and recycling petrodollars.[20] But the Group of 77 countries were not to be dissuaded from concentrating on the concept of global negotiations.

This became clear at the next forum in which the United States made concrete proposals: the UNCTAD IV meeting in Nairobi, Kenya in May 1976. The main issues included the developing countries' debt burden, technology transfer to developing countries, and the establishment of an Integrated Commodity Program (ICP). Secretary of State Kissinger again proposed an International Resources Bank (IRB) to be associated with the World Bank and to encourage private investment in developing countries resources by reducing the risk companies faced. He also advocated an international energy institute. The resolution proposing an ICP passed with even U.S. assent although the United States indicated its affirmative action meant only it was open to additional discussion. The developing countries kept the debate focused on issues of concern to them, while the United States was forced to concede.

In between the next major forum for global negotiations, while the COW was created and CIEC continued its way to a fitful conclusion, the developing countries had other opportunities for organizing their agenda. In August 1976 the Fifth Non-Aligned Summit was held in Colombo, Sri Lanka, and in December 1978 an organizing meeting was sponsored in Arusha, Tanzania for UNCTAD V. The Arusha meeting was particularly important for the sense of unity which came out of the meeting. Previous UNCTAD meetings had floundered because of disunion within the ranks of the developing countries, a failure to articulate a concrete program and to remain unified in the face of Western opposition. The developing countries hoped to enter the UNCTAD V sessions with such a program in place and force the developed countries to capitulate to their demands.

But before the UNCTAD V conference could occur, a major international crisis took place which reemphasized the importance of energy to the global economy. This was the second oil "shock" of 1979 arising from the Iranian revolution's impact on Iran's oil exports. Between December 1978 and December 1979, the price of Saudi market crude doubled.

In March 1979 Secretary of State Cyrus Vance gave a major speech outlining "America's Commitment to Third World Development."[21] The speech presented the U.S. policy for the forthcoming UNCTAD V. As such, it was an attempt to place the Third World's objectives in a context which was quintessentially American, separating the problem into its discrete parts and solving each one individually. It was in much the same format as CIEC. The Secretary of State stated:

> Thus, while we will continue to work with the developing countries in addressing the future of the international political and economic system, we intend increasingly to concentrate on specific development goals: energy; food; health; and increasing the capacity of developing countries to obtain and apply the knowledge and technology they need.

These issues formed the framework for the U.S. bargaining position at the May-June 1979 meeting of UNCTAD V in Manila, the Philippines, but they did not explicitly form part of the prepared agenda.

The provisional agenda adopted by UNCTAD's Trade and Development Board on March 27, 1979, concentrated on 11 issues: (1) developments in international trade (protectionism and multilateral trade negotiations); (2) commodities (the Integrated Program for Commodities); (3) manufactures and semi-manufactures; (4) monetary and financial issues (i.e., increasing aid to developing countries); (5) technology transfer; (6) shipping; (7) the least developed among the developing countries; (8) landlocked and island developing countries; (9) trade and aid with Communist bloc countries; (10) economic cooperation among developing countries; and (11) institutional issues.[22]

Although the UN Secretary-General Dr. Kurt Waldheim opened the meeting stating that energy and food were the critical development issues, a spirit of confrontation held throughout the sessions. The United States concentrated on food, energy, the poor, and technology, while the developing countries reiterated their need for an increase in aid, improvement in the terms of aid, establishment of an independent debt commission, and technology transfer. The developing countries accused the United States of trying to divide the Third World, using energy as a wedge, separating the developing countries into the "haves" (energy-rich states) and the "have-nots" (energy-poor states). The international petroleum market contributed to this. Almost six years from the last oil crisis and during the second, many in the Third World were questioning the wisdom of continued adherence to a unified negotiating platform, particularly since the OPEC nations had been parsimonious in their aid. Despite continued unanimity within Third World ranks during the meeting, no progress was made on any of their issues. The United States continued to refuse to support global negotiations or any resolution featuring interdependence unless energy was mentioned. The Third World would not concede this point, fearing it would insult OPEC. UNCTAD V was a failure.

UNCTAD V was followed shortly by the annual summit meeting of leading industrial nations. On June 28, 1979, the group met in Tokyo, just as OPEC raised its oil price again. Primary focus of the summit was energy. The summit leaders agreed to reduce consumption, setting targets for future maximum demand. President Jimmy Carter returned home to make a major speech on the need to confront the energy crisis as the "moral equivalent of war." But while the West viewed energy as an essential issue of global economic stability, the developing world continued with its own specialized agenda. International meetings later in the summer made this conclusion more apparent.

In July 1979 a World Conference on Agrarian Reform and Rural Development (WCARRD) was held in Rome to develop a program for those issues. Again the United States wanted attendees to address issues of energy, food, health, development capacity, and institutional reforms.[23] In August a UN Conference on Science and Technology for Development (UNCSTD) met in Vienna. Main issues included promoting access to technology and a code of conduct for multinational corporations in developing countries. While only tangentially related to the energy issue, it was a further illustration of U.S. isolation. The conference was at the brink of collapse until the United States agreed to the establishment of a new fund and organization to promote science and technology. Even so, neither the United States nor OPEC was willing to commit money to the new fund.[24]

The frustrations of almost ten years of trying to force the industrial countries to the bargaining table for global negotiations on NIEO culminated in the dramatic statements of the Sixth Non-Aligned Summit Meeting in Havana in

September 1979. Although the Cuban site dictated in part the outcome--Fidel Castro, as leader of the host country, presided--the hostility toward the United States was genuine and in the words of one analyst, "a stark retort to the illusions of the United States, exemplified at WCARRD and UNCSTD, that it can win the hearts and minds of the Third World with good intentions or even cash."[25] The conference supported OPEC's position and attacked the West for using the energy issue as a means of fragmenting Third World unity. The world's economic problems, including inflation, were blamed on the West and its multinational corporations.

This bitter meeting was followed by an even more acrimonious one in New Delhi from January 21 through February 9, 1980, of UNIDO III.[26] Totally opposing views were taken by the developed and developing countries on energy, technology, finance, investment, and multinational corporations. Emboldened perhaps by their successes in Havana, the radicals of the developing world sought confrontation rather than cooperation.[27]

However, while relations between the North and the South were at their most vituperative, changes were occurring to suggest a reconciliation of relations. In June 1980 another summit of industrial nations was held in Venice. Again, energy was prominent on the agenda. Talk focused on how developed countries could sever the link between oil and economic growth, a link which suggested increasing oil demand was inevitable. In addition, corollary issues relating to energy were raised: oil price stability and petrodollar recycling. It was agreed to ask the World Bank to evaluate the adequacy of energy resources in the Third World, and the final summit statement stated that they were looking forward to future global negotiations.

The frustrations of the decade seemed to be working themselves out. At the Eleventh Special Session of the United Nations in August 1980, discussions were held on how to reorient the international economy. On September 24 at the United Nations, Secretary of State Edmund Muskie labelled energy as the central question. The United States appeared to acknowledge the need for global negotiations for an international economic order, while the Third World, seemingly spent after its latest round of rhetoric, acknowledged the importance of energy and other issues to development policy. But the atmosphere of compromise was altered with the new U.S. administration.

The Reagan Administration, the most ideological to take office in a good number of years, was eager to challenge the Third World. It began by rejecting almost out of hand the proposed World Bank energy affiliate. The private sector would be the main vehicle for achieving U.S. foreign economic policy goals to the detriment of multilateral or even bilateral aid programs. President Ronald Reagan proposed a new Caribbean Basin Initiative (CBI) to aid the developing nations in that region while simultaneously slashing the U.S. contribution to the "soft loan" window of the World Bank, the International Development Association

(IDA). The President explained this reduction by insisting that "the most important contribution any country can make to world development is to pursue sound economic policies at home."[28] By concentrating on the United States' economic woes first, the President ultimately hoped to benefit others. This drastic shift in approach terrified U.S. allies, as became clear during the annual economic summit in Ottawa, Canada. Both the 1981 and 1982 summits resulted in confrontations between the new U.S. administration and its allies. Domestic economic policies in the United States were having a major effect on the world, plunging it into a recession. The United States had East-West issues and loosening trade restrictions on its agenda while its allies were concerned about high U.S. interest rates and exchange rate policy. This remarkable consistency of concerns continued into the third summit at Williamsburg, Virginia in May 1983. At each of these meetings the United States opposed any reference to "global negotiations" in the final communiqué but capitulated at the end as a sign of accommodation.[29]

At the July 19-21, 1981 Ottawa summit, President Reagan wanted talks to focus on East-West issues although the previous year's summit in Venice had placed North-South problems high on the West's agenda.[30] This simple dichotomy characterizes best the U.S. approach under its new President. Problems of the Third World, including energy, were to be viewed through this prism. The first test of the new administration's resolve and ability to determine the direction of the debate was the August 1981 UN Conference on New and Renewable Sources of Energy (UNCNRSE) meeting in Nairobi and the October 1981 special 22-nation North-South "survival" summit in Cancun, Mexico. The Reagan approach was to appear accommodating while being intransigent. The UNCNRSE was approved by the UN General Assembly in 1978 but was not held until August 1981. The purpose of the conference was to: (1) create an awareness of the energy problem, (2) examine the possible contributions of new and renewable sources of energy to world energy supplies, and (3) identify "areas of action."

The UNCNRSE focused on one of the energy areas which the Reagan Administration was on record as opposing: government investment in new and renewable sources of energy. In its view market forces should determine the U.S. and the world energy mix.

The nervousness of the West, especially the United States, about attending a conference which might address more substantive "political" issues rather than purely technical ones was clear. But despite the declaration of the Conference's Secretary-General that the meeting was "no more and no less that," it was difficult to avoid such political issues.[31]

The Program of Action adopted in Nairobi called for: (1) the establishment of an intergovernmental body in the United Nations to deal with new and renewable sources of energy and to monitor and implement the Program of

Action--in effect creating a new UN organization; the West as a whole, however, opposed the establishment of a separate UN energy institution; instead, the West wanted the role of the UN Committee on Natural Resources expanded; and (2) recommended specific measures for programs to develop renewable resources and called for the establishment of energy information centers which would be joined in a global information network.

While the official U.S. response was that the conference reached "a successful conclusion" and that "this was an unusually successful meeting by UN mega-conference standards,"[32] others charged that the United States "stonewalled" the "opportunity for cooperation."[33]

> The U.S. implacably opposed increases in multilateral energy aid, creation of a World Bank energy affiliate or chartering of any new agency that would funnel aid or information to those nations that desire it . . . the U.S. opposed targets or goals for international renewable energy programs and sought to vest all authority for any international programs in the tightest controlled UN natural resources committee.[34]

Following the session, an Interim Committee met in June 1982 to discuss the Program of Action, but agreement could not be reached on financial or institutional questions, and they were brought up in the General Assembly in December. At that time a vote of the Assembly approved a resolution to establish an Intergovernmental Committee on the Development and Utilization of New and Renewable Sources of Energy although it was opposed by both the United States and the Soviet bloc countries on budgetary and organizational grounds. The same resolution also reaffirmed the importance of a World Bank energy affiliate. A second resolution was approved by every country but the United States, again for budgetary reasons, that called on the Secretary-General to report on the development of energy resources in developing countries and review their energy investment needs. A follow-up meeting was held in Rome, June 7-18, 1982 to implement the Program, but no agreement could be attained on the key issues of financial and institutional arrangements. The Interim Committee on New and Renewable Sources of Energy only reached agreement to recommend that a consultative mechanism be established for further discussion.

In effect the Third World accomplished many of its political objectives for the UNCNRSE by bypassing the committee and conference structure and proposing resolutions in the plenary sessions of the United Nations General Assembly. This, however, could not have much impact without budgetary allocations. And these allocations, dependent as they were on funding from the industrial states, did not occur.

The failure to achieve financial support in Nairobi for new institutions was matched at Cancun, Mexico, in October

1981 by the failure of the developing countries to obtain substantive political concessions from the industrial West. The purpose of the Cancun Summit at which 22 nations represented the North and the South was to "forge a common approach by industrial nations and developing countries" on growth.[35] But the U.S. objective was to "reverse the sweeping demands of developing countries over the past decade."[36] In this the United States was largely successful.

The United States managed to appear accommodating while forcing everyone to agree to its demands. The U.S. negotiating approach was to play the classic interrogator's ploy of "good guy-bad guy." On one hand, policy makers made soothing speeches about the importance of development. For example, before the United Nations on September 21, Secretary of State Alexander Haig outlined three areas for international action: a global expansion of trade, an increase in investment, and stronger international cooperation in food and energy. In a major address in Philadelphia just before Cancun, President Reagan stated:

> We will support energy lending by multilateral institutions provided the projects are economically viable and they expand developing country energy production through greater private investment.[37]

On the other hand, both before and during the meeting, U.S. negotiators took an intransigent position. At the preliminary talks, the United States argued over language: It was willing to "discuss" but would not "negotiate" global issues. It opposed any proposal to increase aid to a certain benchmark percentage because that would be "a form of international taxation without a structured institution."[38] The United States opposed a formal agenda, wanting the summit to be "as informal as possible, without a preparatory document, fixed or final communique."[39]

The United States was partly successfully because the Third World saw it as an opportunity to "sensitize" President Reagan to Third World problems. They needed the United States President there to do that. Also, it was a small meeting; only a few representative countries attended. Without the United States, therefore, the meeting could have had no impact. The United States also played it for the public relations gimmick it really was. At the conclusion, the United States dramatically agreed to "global negotiations" but only if they followed certain conditions:

(1) that they be held in an existing organization, e.g., IMF, General Agreement on Tariffs and Trade (GATT), or the World Bank;
(2) that they focus on specific proposals;
(3) that they not involve massive resource transfers; and
(4) that they be undertaken in a cooperative spirit.

In addition, the United States agreed to holding talks in the United Nations as long as nothing was voted on. And finally, the United States did not agree to talks per se but to continue "preparations" for such talks.[40]

Of course, the United States was not alone in its views on the issues discussed at Cancun--raw materials, trade, energy, aid, and finances. On trade, for example, most of the industrial nations wanted the developing countries to join GATT. And in energy, while some countries wanted stable prices, others did not. Also, even though the United Kingdom agreed to support a World Bank energy affiliate, both OPEC nations and the United States opposed it--the former because they feared it would undermine their influence, and the latter because it wanted the private sector to provide the financing.[41]

But even the agreement to hold further talks led nowhere. The "spirit of Cancun" was soon dead. An attempt to restart global negotiations was prevented by the United States in December, and an effort in April 1982 to circumvent the U.S. conditions was also stopped.[42] A classified report by the Central Intelligence Agency especially commissioned by the U.S. Ambassador to the United Nations, Jean Kirkpatrick, purporting to show that global negotiations would undercut the independence of the World Bank and IMF, was leaked.[43]

The United States continued to pay lip service to the interest of its industrial allies that it appear accommodating to developing coountry demands while supporting no new initiatives. The June 6, 1982, Versailles economic summit communiqué stated that the "launching of global negotiations is a major political objective approved by all participants in the summit." At this meeting of major industrial nations, specific mention was made of the importance of food and energy aid to developing countries and support for IDA. A second case in point is the Reagan Administration's failure to support financially World Bank energy lending while agreeing to the concept in principle.

The World Bank's energy lending had concentrated on electric power, with some natural gas lending. After 1976, the Bank became more involved in petroleum development. Discussions during CIEC and elsewhere had focused on the concept of establishing within the World Bank Group an affiliate for energy lending, especially for petroleum development, much as the International Finance Corporation (IFC) operated. The first staff study was proposed to the Bank's Board of Directors in 1977. In July they approved financing for oil and gas on an experimental basis but not for exploration. Annual energy lending would have increased to $500 million by 1982.

The proposal was discussed at the 1978 Bonn economic summit. President Carter had earlier approved a general increase in the Bank's funding, but opposition to an expanded Bank energy lending program soon surfaced both within the government and outside. Parts of the U.S. Departments of Energy, Treasury, and State opposed the program for a variety of reasons:

(1) an exploration fund would be uneconomic and risky;
(2) no clear proof had been given that a capital shortage restrained developing countries exploration; political risk and market conditions were seen as more important factors; and
(3) Congress would oppose additional Bank funding.[44]

At the economic summit the proposal was tabled for further study although the United States supported an expansion of energy assistance. In January 1979 the U.S. representative to the Bank voted to expand the existing Bank program into natural gas and petroleum. However, much of the U.S. energy industry, especially Exxon, opposed the Bank program as unwarranted interference in private sector initiatives. Despite this opposition, the Bank proceeded to formulate a program for establishing a separate energy affiliate. A study was released by the Bank in August 1980, entitled "Energy in the Developing Countries," which proposed a study of the energy affiliate concept. The study, however, was already well under way and in the spring of 1981 Robert MacNamara, President of the World Bank, formally submitted the recommendation. It was quickly rejected by the Reagan Administration after the most facile examination, declaring that "if appropriate terms and conditions are offered, private companies can be expected to provide the bulk of the investment resources for developing countries oil and gas development."[45]

The same inconclusiveness was shown in the Cancun "survival" summit in 1981. The representative developing countries argued for global negotiations to establish new levels of aid and a change in the voting arrangements of the major multilateral trade and financial institutions (IMF, World Bank, and GATT) in which they held a minority voice. This the West opposed although it did suggest that discussions on an NIEO could be held within the framework of existing institutions. Although the meeting was hailed as a success because world leaders were able to discuss serious development issues, "the leaders' only firm decision was that they had no mandate to decide anything."[46]

Thus, despite a decade during which energy and the effects of energy on the world's economy were the dominant themes, the world's leaders were unable to develop a systematic and pragmatic approach to solving energy problems. In part, the failure for direction must lie with the political debates between the developed and developing countries. The Third World did not want attention diverted from the comprehensive package of international reforms that they had developed in NIEO, while the developed countries were eager to pursue, what was to their minds, more concrete, short-run solutions which incidentally and not unwelcomingly would fragment the Third World's international political power. The result of this struggle was a stalemate. The West held the financial power necessary to implement any program. Yet, as the debate progressed, there was a gradual weakening on both sides. After all, some of the issues of concern to the Third World directly affected the well-being of the West: international debt, exchange rate fluctuation, commodity prices, and technology transfer. At the same time the Third World had come to realize that OPEC was not in a position to lead it out of poverty, especially as oil prices weakened. Whether or not recognition of the interdependence in the world economy will result in new joint solutions to the world's problems, however, is not now guaranteed.

U.S. intransigency has in part worked. Rather than leading to new rounds of inflammatory statements about U.S. imperialism and further deepening the cleavage between North and South, the developing countries appeared to become more accommodating rather than less. The Seventh Non-Aligned Summit meeting in New Delhi in March 1983 "was unusually moderate, and more pleading than condemning, toward the West." The issue of global negotiations was put on the back shelf, and for the first time "immediate measures," such as an enlarged IMF lending program, were emphasized rather than sweeping international reforms.[47]

## CONCLUSIONS

The last three decades have seen an evolution in the attitudes of states to managing their energy resources. Early attempts to assert national sovereignty over resources failed, but during the 1960s and the 1970s, inspired by OPEC's leadership, individual states were able to wrest control over petroleum development from the international oil companies. The 1980s is a period of reestablishing a balance based on interdependence. Energy demand has proven to be highly elastic in the long term, undermining cartel behavior. But the objectives of cartel behavior remain: maximum revenues without wide price fluctuations. In the 1980s, both oil consumers and oil producers began to recognize the benefits of price stabilization policies to a world economy which had become highly integrated.

In the past main issues have been financial resource transfers, physical resource controls, and support for resource development. The West has fought every effort by the developing countries to attain their objectives in this area. The West has opposed debt forgiveness and renegotiation, opposed nationalization and expropriation, and opposed new programs such as the World Bank's energy affiliate. But the West has had to give ground in each of these areas. Third World debt has had to be renegotiated, nationalization accepted, and multilateral energy programs expanded. Because of economic interdependence, the West has won many battles but may have lost a war.

## NOTES

1. Hearings before the Subcommittee on Energy of the Joint Economic Committee (JEC), U.S. Foreign Energy Policy, 94th Cong., 1st sess. (1975), Washington, D.C.: Government Printing Office, 1976, p. 19. (Citing testimony by James Grant.)

2. Somendu Kumar Banerjee, "The Concept of Permanent Sovereignty over Natural Resources--An Analysis," Indian Journal of International Law 8 (October 1968):518.

3. Operative paragraph 4, United Nations General Assembly Resolution 1803 (XVII) of December 14, 1962.

4. The 1960s was a period of relative economic prosperity in the world. The problems of the developing nations were oppressive but not viewed as impossible to resolve. The fact that in 1964 developing countries could convince developed countries to attend a conference on trade and development (UNCTAD I) was an indicator of relative prosperity rather than a measure of despair. Both sides could afford to demand and receive some concessions from the other without real sacrifice.

5. See C. Fred Bergsten, "The Threat from the Third World," Foreign Policy 11 (Summer 1973):102-24.

6. Penelope Hartland-Thunberg, "The Pervasive and Continuing World Economic 'Shock' from the 1979-80 Oil Price Increase," paper presented at a Conference on Persistent Oil Vulnerability, Center for Strategic and International Studies, Georgetown University, Washington, D.C., December 8, 1982, p. 5.

7. JEC, Foreign Energy Policy.

8. Hearings before the Senate Subcommittee on International Economic Policy, The U.S. Stake in the Global Economy, Washington, D.C.: Government Printing Office, 1981.

9. JEC, Foreign Energy Policy, p. 59.

10. JEC, Foreign Energy Policy, p. 19.

11. Robert F. Meagher, An International Redistribution of Wealth and Power, A Study of the Charter of Economic Rights and Duties of States (New York: Pergamon Press, 1979), p. 117.

12. JEC, Foreign Energy Policy, p. 60.

13. New York Times, December 13, 1973, p. 29.

14. Meagher, International Redistribution, pp. 93-94.

15. Meagher, International Redistribution, pp. 94-95.

16. JEC, Foreign Energy Policy, p. 62.

17. David Adamson, "Multilateral Energy Negotiations: Lessons of the Conference on International Economic Cooperation" (unpublished paper, August 1981), p. 12.

18. Ibid., p. 4.

19. Meagher, International Redistribution, p. 128.

20. Robert K. Olson, U.S. Foreign Policy and the New International Economic Order, Negotiating Global Problems, 1974-1981 (Boulder: Westview Press, 1981), p. 105.

21. Speech delivered by Secretary Vance at the Northwest Regional Conference on the Emerging International Order, Seattle, WA, Department of State Bulletin, March 30, 1979.

22. See UNCTAD Document Td/220. "Provisional Agenda for the Fifth Session of the United Nations Conference on Trade and Development," April 12, 1979

23. Olson, U.S. Foreign Policy, p. 80.

24. Michael C. Oppenheimer et al., Prospects and Policy Initiatives for the United Nations Conference on New and Renewable Sources of Energy: A Report to the U.S. Coordinator (Glastonbury, CT: The Futures Group, January 1981), Report #493-34-03, pp. 4-5.

25. Olson, U.S. Foreign Policy, p. 87.

26. Oppenheimer, Prospects and Policy Initiatives.

27. Olson, U.S. Foreign Policy, p. 103.

28. Address by President Reagan before the annual meeting of the IMF, IBRD, and IFC, "Challenge of World Development," Washington, D.C., September 29, 1981.

29. Steven R. Weisman, "Accord Viewed as Symbolic, with Deep Rifts Remaining," New York Times, July 22, 1981, p. D18.

30. Washington Post, July 9, 1981, p. A14.

31. UNCNRSE, Conference News, "Secretary-General Iglesias Outlines UNCNRSE Targets," No. 9, p. 1 (no date).

32. Hearings before the House Subcommittee on Energy Conservation and Power, October 21, 1981 (Testimony by Elliot Abrams).

33. Ibid. (Testimony by Erik Eckholm).

34. Ibid. (Testimony by Jim Harding).

35. "Britain Deals U.S. a Surprise Setback at Cancun Meeting," Wall Street Journal, October 22, 1981, p. 2.

36. Art Pine, "Reagan Makes Conditional Offer to Join 'Global Negotiations' on Third World Aid,'" New York Times, October 23, 1981, p. 4.

37. Clyde H. Farnsworth, "Reagan Bids Third World Adopt Free Enterprise to Battle Poverty," New York Times, October 16, 1981, p. A12.

38. Bernard D. Nossiter, "Haig Aide Says U.S. Opposes Sharing the Wealth," New York Times, August 6, 1981, p. A6.

39. Alan Riding, "Preliminary Contacts Sought for North-South Conference," New York Times, August 7, 1981, p. A6. Interestingly, the United States followed the same approach with its allies at the 1983 Williamsburg Summit.

40. Pine, "Reagan Makes Conditional Offer."

41. "What the Poor Will Ask," New York Times, October 18, 1981.

42. Art Pine, "The Promise of Cancun Is Vanishing as Rich and Poor Nations Bicker Anew," Wall Street Journal, April 12, 1982, p. 26.

43. Seth Lipsky, "The U.S. and Global Negotiations," Wall Street Journal, December 16, 1981, p. 26.

44. U.S. General Accounting Office (GAO), Formulations of U.S. International Energy Policies, ID-80-21, Washington, D.C., September 30, 1980, pp. 37-41.

45. U.S. Department of Treasury, Office of the Assistant Secretary of International Affairs, An Examination of the World Bank Energy Lending Program, Washington, D.C., July 28, 1981, p. ii.

46. Richard Nations, "Pragmatism at Cancun," Far Eastern Economic Review, October 30, 1981, p. 59.

47. Paul A. Gigot, "Third World's Stance Softens Toward the West," Wall Street Journal, March 14, 1983, p. 36.

# PART II

## THE UNITED STATES AND THE MULTILATERAL MANAGEMENT OF FOOD RESOURCES

# 3

# AMERICAN MULTILATERAL FOOD DIPLOMACY: THE UNITED STATES ROLE IN INTERNATIONAL ORGANIZATIONS

*Raymond F. Hopkins*
*Donald J. Puchala*

The United States participates in a myriad of international organizations concerned with problems of the global food system. Eighty-nine international intergovernmental bodies were recently listed in a report on U.S. organizational activities in world food affairs prepared by the United States Senate.[1] If one were also to count the private associations that include U.S. members, our multilateral involvements in international food trade, aid, research, education, and communication would number in the thousands. Needless to say, it would be a large task, well beyond the scope or intent of this chapter, even to begin to map the full structure of the international organizational arena for food, and even lengthier efforts would be required to assess in any detail the specific operations and impacts of these countless bodies. While we do present some broad structural mapping and organizational evaluation, the main point of this chapter is to explore the relationship between U.S. interests and past and future U.S. participation in international organizations dealing with food questions. In this regard, we shall be paying special attention to U.S. dealings with the United Nations Food and Agricultural Organization (FAO). The most important recommendation that emerges from this analy-

---

This chapter has been extracted from Raymond F. Hopkins and Donald J. Puchala, Global Food Interdependence: Challenge to American Foreign Policy (New York: Columbia University Press, 1980), pp. 127-149. Reprinted by permission.

sis is an urging toward more active, coordinated, and positive U.S. participation in food organizations.

## INTERNATIONAL FOOD ORGANIZATIONS

With the special exception of organizations instituted to monitor and regulate international dealings in particular agricultural commodities, and the notable exception of the General Agreement on Tariffs and Trade (GATT), the most important international organizations involved in food affairs are all elements of the United Nations system. Central among these is the United Nations Food and Agricultural Organization. FAO is an autonomous association, linked to the United Nations system by agreement between the FAO Conference and the General Assembly in 1945. It is responsible only to its members (currently 136 national governments) and is financed by them for its regular budget, which largely pays for headquarters costs and staff operations.[2] The charter mandate of FAO calls upon it: (1) to collect, analyze, and disseminate information relating to food and agriculture; (2) to provide an international forum for consideration of food problems; and (3) to provide technical assistance to member countries, all with the ultimate objective of "ensuring humanity's freedom from hunger."[3]

Most closely bound to FAO (or vice versa) within the UN system is the United Nations Development Program (UNDP) (See Figure 3-1.). The main source of technical assistance in the UN, UNDP was founded in 1966 through a merging of the Expanded Program of Technical Assistance (EPTA) and the United Nations Special Fund. FAO currently serves as the executing agency for most projects in food and agricultural development financed by UNDP, and UNDP reimbursements yearly constitute the single largest category of FAO's total receipts--about double that of the regular budget.[4]

Less intimately associated with FAO, but still importantly linked, are the World Bank Group, the three regional development banks, the World Food Program, and other "cooperative" programs. Through a World Bank/FAO Program, FAO assists the Bank in identifying and evaluating projects for possible funding, and it aids prospective loan recipients in preparing applications. Less direct links have existed between FAO's Industry Cooperative Program, its Investment Center, and the Bank's International Finance Corporation, where liaison and exchanged information guide private-sector investments in food processing and agricultural development. By way of their capital-raising activities, and via intraprofessional communication, the World Bank Group is connected to the Inter-American Development Bank (IDB), the Asian Development Bank (ADB), and the African Development Bank (AFDB), all of which finance agricultural development.

The World Food Program (WFP), created in 1962, is essentially an international food-aid operation that principally finances projects, funding, and development with food and modest overhead support pledged by members. WFP also intermittently functions to supply disaster and famine relief (about 10 percent to 15 percent of its outlay), working with the UN Disaster Relief Office (UNDRO). Organizationally, it is linked to FAO via appointments to its governing body, via cooperative field operations, and via having its main offices at FAO headquarters, where FAO provides heavy staff support (for which WFP reimburses FAO). Cooperative programs also link FAO to various other parts of the UN system; these include a program with the World Health Organization (WHO) on food standards (the Codex Alimentarius Commission), a program with the United Nations Children's Fund (UNICEF) directed toward improving nutrition among children, and a program in education about global hunger and food needs, the Freedom From Hunger Campaign.

Several new organs were created by the World Food Conference in November 1974, and these are now operating in the UN food network. Resolutions approved first by the Conference and later by the UN General Assembly established a World Food Council (WFC) as the "highest" institution on world food problems in the UN system. Meeting at the ministerial level, the WFC is composed of 36 countries, nominated by the Economic and Social Council (ECOSOC) and responsible to it.[5] Its composition represents a world fragmented into an industrialized North, a developing South, a capitalist West, and a socialist East. For broad policy issues, the WFC was intended to be the hub of the UN food network. Whether it will achieve this status remains to be seen. After an initial conflict over personnel and budget for operations, the WFC has had two meetings (1977 and 1978) that elevated substance above rhetoric. The assessment of progress at its Fourth Annual Meeting in Mexico City, June 1978, suggested that many of the goals of the Rome Food Conference (1974), which the WFC was to shepherd into existence, had not been realized in part or at all.[6] According to conference reports, the 10 million ton minimum food-aid goal was barely reached, the grain reserve seemed unlikely to succeed, and many other steps to alleviate hunger had not reduced the number of people who eat too little--well below minimum standards. Since 1978 the WFC Secretariat has put most of its lobbying and coordinating capability into efforts to improve poor-country food supplies through rural development. Its clearest success was in the adoption of national food system studies as agreed to and funded at its 1979 meeting in Ottawa.

Also authorized at Rome in November 1974 was the International Fund for Agricultural Development (IFAD). After an initial capitalization of just over $1 billion had been raised, the IFAD began operations in 1977 with the special aim of helping the poorest of the developing countries. It reports to, but is not an authority under, the Secretary General of the UN, and its operations are overseen by a governing board composed of representatives of three categories of countries--developed donor, developing donor (e.g., OPEC), and developing recipient.[7] It has a complex

Figure 3.1
The United Nations Food Network: International Organizations in Food and Agricultural Development (Partial Listing)

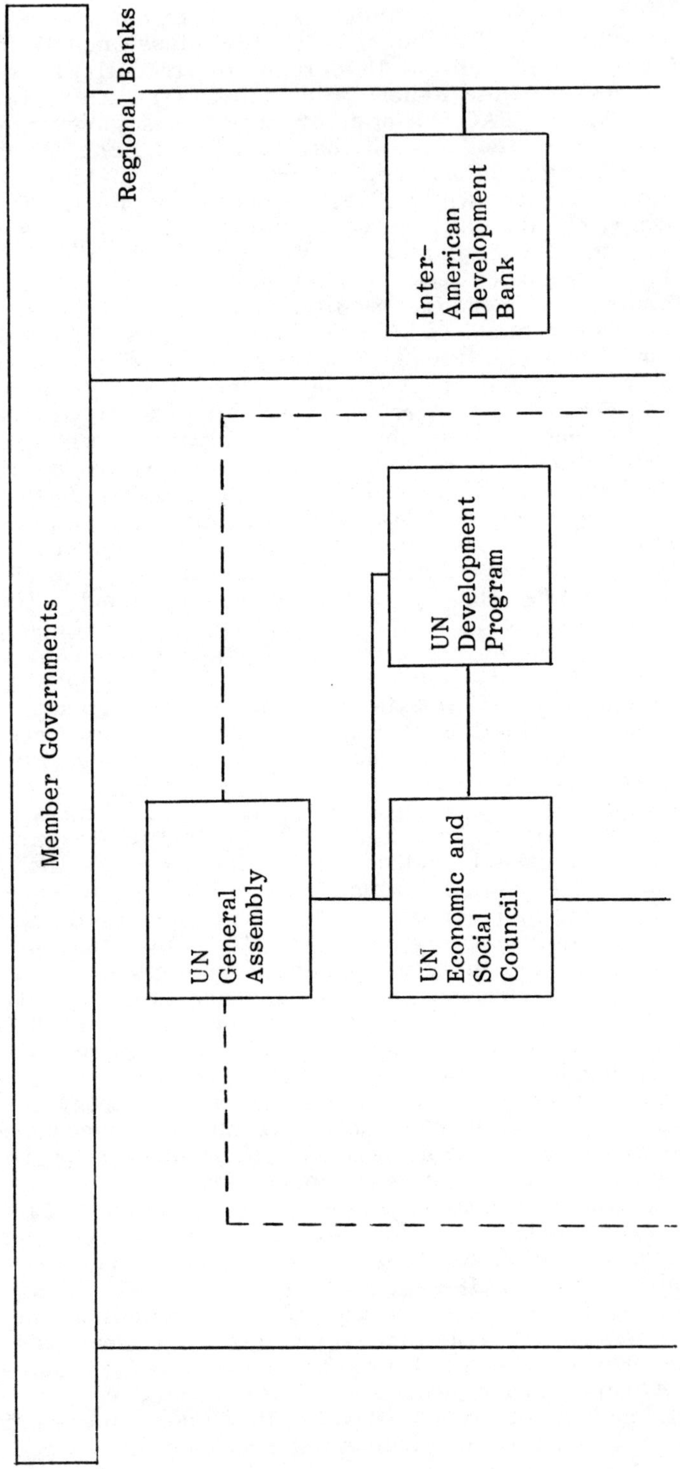

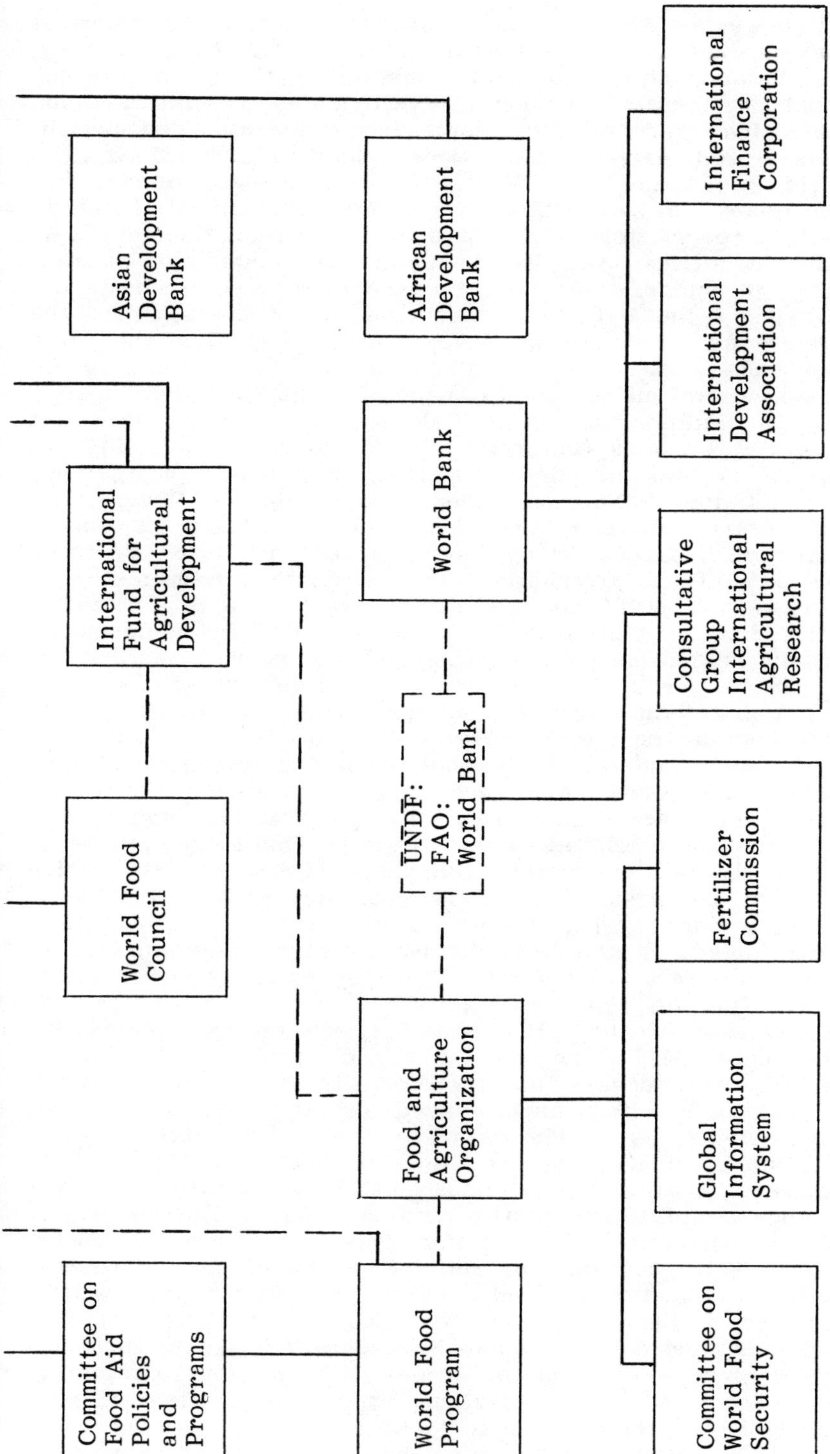
Asian Development Bank
African Development Bank
International Finance Corporation
International Fund for Agricultural Development
World Bank
International Development Association
Consultative Group International Agricultural Research
UNDP: FAO: World Bank
World Food Council
Fertilizer Commission
Food and Agriculture Organization
Global Information System
Committee on Food Aid Policies and Programs
World Food Program
Committee on World Food Security

voting system and thus far relies on the major development banks to assist in project appraisals.

Established in 1969, the Consultative Group on International Agricultural Research (CGIAR) is an international consortium of donor and recipient governments, development banks, and private foundations, sponsored by FAO, the World Bank, and the UNDP. It functions to support and coordinate the activities of eleven agricultural research centers specializing in problems of food production in developing countries, and it administers a budget for them of over $80 million.[8] As with the IFAD, representation to the CGIAR by the U.S. government is under the aegis of the Agency for International Development (AID), but the State Department, Treasury, and Agriculture all expect to be consulted and all attempt to shape U.S. participation.

In addition to organs already described, there are several UN bodies that regularly consider food questions as aspects of their broader programs. Significant among these is the United Nations Conference on Trade and Development (UNCTAD), where intense debate has centered on North-South trade issues, many having to do with terms of trade for agricultural commodities and commodity agreements.[9] In many ways, UNCTAD has become as salient a forum for North-South agricultural issues as has GATT for developed Western states bargaining on agricultural issues.

Some words of caution are in order before we move from this quick "mapping" into analysis and evaluation. First, organizations depicted in Figure 3-1 and briefly highlighted here are by no means a complete list of international food bodies, not even as regards the UN network. For one thing, there are various coordinating committees and ad hoc food groups interlaced among the major institutions, such as the Committee on Surplus Disposal (CSD) of FAO which monitors concessional sales for possible violations of anti-dumping norms. For another, many of the major organizations noted, FAO in particular, contain any number of quasi-autonomous, differentially responsible organs within them. Second, the operational world of the groups and associations of the UN system is not nearly as orderly, well-organized, or separate as the discussion suggests. In reality, redundancy (for good or ill) is rampant, complementarity is often unrecognized or at least unexploited, responsibility is poorly defined as is accountability, coordination is difficult, and political and bureaucratic activities buttress institutional autonomy and proliferation to add further complications to the network. Third, let us caution against mistaking activity for impact or accomplishment. There is a good deal of activity surrounding the international organization of food affairs, but budgets are modest, authority is limited, support from member states is tentative, and, for myriad political and bureaucratic reasons, including a strong imperative not to tamper with the sovereign integrity of any state, organizations tend to be restrained from accomplishing their mandated tasks.[10]

U.S. officials and political leaders in recent years have increasingly found international organizations inhospitable.

The voting majorities for U.S. preferences in the United Nations have long eroded, actions against Israel and South Africa and other issues have caused the United States to leave the International Labor Organization, to threaten to abandon UNESCO, and to complain about the strident and unproductive rhetoric in many other forums. This skepticism has been less for FAO than for other bodies, but the United States has nonetheless sought to avoid the strongest demands arising in these meetings and has encouraged smaller groups to work out international food questions whenever possible.[11] U.S. preference for moving the discussion of reserves to London and to the International Wheat Council is a case in point.

## INTERNATIONAL ORGANIZATIONAL FUNCTIONS AND FOOD AFFAIRS

It is, however, generally ill-advised for U.S. leaders to evaluate international institutions in terms of unrealistic and historical criteria. Many global food problems could be more effectively addressed if international organizations were more authoritative, more efficient, more respected in the eyes of member governments, more preferred as diplomatic arenas, and more able to command the resources required for fulfilling their objectives. Yet, rather than dwell upon functions international organizations do not, and perhaps cannot, perform, or perform ideally, given the environment in which they operate, note should be made of the functions they can and do perform in international food affairs. It is in this "realistic" analysis of the functional context of international organizations that we can best evaluate the advantages of U.S. participation.

In our estimation, at least five important functions in world food affairs are well served by existing international organizations. They are each deserving of U.S. encouragement and support.

## INTERNATIONAL ORGANIZATIONS AS LOCATIONS FOR CONSULTATION AND INTERNATIONAL EDUCATION

The congresses, conferences, and committees of international organizations are first and foremost meeting places where national leaders or their representatives can come together to exchange views. If they did not exist, imperatives for international communication would invariably create them. Forums provided by international food institutions range from massive international conferences such as the World Food Conference in November 1974, or the biennial conferences of FAO, which set themes and tones for subsequent food diplomacy, to small seminars for experts where detailed questions are debated and technical information is exchanged. In fiscal 1975, for example, U.S. officials attended 77 international organizational meetings related to food and agriculture. More than 240 U.S. delegates were involved at a cost of approximately $340,000.[12] Estimated costs incurred in preparing for such international meetings were in the neighborhood of $1.6 million.[13] Most involved in these international forums deemed them important and worth the costs in time and money they entailed. For example, when asked to cite benefits to the United States from participation in FAO, former Assistant Secretary of Agriculture Richard Bell told a Senate subcommittee that the first benefit was in

> providing forums [in which] agricultural, fisheries, food and forestry policy and technical problems are discussed with representatives of other countries. . . . Through them it [the United States] obtains a great deal of useful information, develops better understanding or agreement with other countries on problems of common concern and contributes to improved international relations generally.[14]

This view was also stressed during the same hearings by Samuel W. Lewis, former Assistant Secretary of State for International Organization Affairs:

> On the benefit side, the greatly increased [FAO] membership provided a forum for balanced consideration, in most cases of problems which too often in the past had been considered in isolation. A global economy becoming steadily more interdependent made such dialogue indispensable. FAO sessions, particularly technical meetings or seminars in developing countries, have provided many opportunities for exerting United States influence on a large variety of agricultural matters of interest to us.[15]

There is no clear evidence that organizational forums enhance international cooperation, that they render issues more tractable than they might be otherwise, or that they facilitate agreement more readily than bilateral dealings. In fact, the contrary is probably true, inasmuch as discussions within international organizations tend to politicize technical issues, complicate simple ones, and unaccountably simplify complicated ones. Nevertheless, international discussion in organization forums does raise consciousness about the gravity of issues, and it exposes all governments to the views, perspectives, and logics of all others. As such, it promotes international insight. In addition, interactions in organizational arenas generate information essential to national policy making, and the emphases and urgencies exhibited at international gatherings directly influence national agendas and focuses of policy attention. Furthermore, since the governments of a great many developing countries currently prefer to conduct most of their international relations via multilateral organizations, business carried on in preparation for, during, and in response to their meetings and conferences is bound to become an increasingly large proportion of the total substance of international relations. What is generally true in this regard for global affairs will be particularly true for food and agriculture affairs, since this area is already one of the most highly organized internationally.

### Initiative and Prodding

Through reports, multilateral conferences, speeches by officers, good offices, and other means of quiet diplomacy, international organizations often persuade or prod member governments to confront issues or consider policy options that national officials might otherwise choose to ignore. Here, there are any number of recent cases in point. Latest impetus for the idea of an international grain reserve, for example, came from a 1973 proposal by former FAO Secretary-General Boerma.[16] The scheme contradicted U.S. and Canadian policies at the time,[17] but the international attention it attracted, and the domestic support it received outside of the U.S. government, prompted reconsideration and review, and Boerma's initiative became one of the elements that brought a change in the U.S. position by 1974.[18] Even more telling as an example of the impacts of international organizational prodding are some of the results of the World Food Conference in 1974. Despite the political polemics and apparent obduracy of the many national positions expressed in Rome, the themes of impending scarcity, food security, and greater self-sufficiency via rural modernization, emphasized in preparatory documents drafted by the FAO staff, subsequently became paramount issues in food diplomacy and important inputs into developing countries' national planning.

### Accumulating and Disseminating Information via Organizational Activities

By collecting information, fostering professional communication, and sponsoring research, international organizations provide valuable services that national governments might either take little interest in or forgo because of the high costs involved. As the principal intergovernmental organization for food affairs, FAO reviews and reports upon nearly all international activities in the field.[19] It also extensively monitors national agricultural conditions as well as the global status of forests and fisheries. Only the research activities and output of the United States Department of Agriculture are in any way comparable to FAO's information programs, as only these two sources comprehensively depict global conditions. Only FAO materials are universally disseminated. FAO's budgets for research and publication are small compared to amounts spent by governments for nationally focused programs. Yet, the organization's output is substantial; its periodicals, yearbooks, and country analyses are frequently cited, and its projections frequently guide national planning and policy making.[20] The network of institutions involved in international agricultural research is centered in FAO, but it extends beyond it to include other public organizations, such as national research and development agencies including the United States Agency for International Development and Canada's International Development Agency (CIDA). It also includes a number of quasi-public bodies, such as the eleven international research centers coordinated by the consultative Group on International Agricultural Research.[21] Widespread international communication about agricultural research is a development of the last two decades, primarily fostered by international organizations. While a full assessment of the institutions in this area is premature, it is fair to credit cooperative international research ventures during the 1960s with producing the "miracle seeds" for high-yielding dwarf varieties of wheat and rice that led to what some proclaimed as a "Green Revolution" in food production.[22]

## Assistance and Advice

International organizations operating in the food and agriculture field have become increasingly involved in funding rural development, technically assisting national governments, and otherwise conducting field operations designed to contribute directly to improving agricultural and nutritional conditions in underdeveloped regions.[23] The list of activities engaged by FAO directly and as agent for the UNDP, by other specialized agencies of the United Nations including the World Health Organization and UNICEF, by UNDRO, the WFP, and the World Bank Group, is much too lengthy to begin to elaborate. However, three points concerning development-assistance efforts on the part of multilateral agencies are important. First, these have been expanding rapidly over the past several years, so much so, in fact, that FAO's budget for field activities now far exceeds its regular budget. The organization, especially at the urging of its current Director-General, Dr. Saouma, has become increasingly a development-assistance institution.[24] Second, development assistance (and other services) performed by international organizations often take the form of activities that national governments will not, or cannot, perform for political reasons. For example, activities deemed interventionist or "imperialistic," if attempted on a bilateral basis, are often acceptable to host governments if attempted by multilateral agencies. Similarly, countries that might be denied assistance on the basis of the ideological concerns, or political priorities, of leading national donors often find international organizations sympathetic to their proposals and needs. Third, in recent years there has been a rethinking and retargeting of aid and technical assistance given by international organizations toward greater emphasis on the interests and needs of small farmers in poor countries.[25] The World Bank in particular has greatly increased the proportion of development assistance lent for improving the lot of the poor through agriculture and rural development assistance. Lending grew from $944 million in fiscal 1974 to $3,270 million in 1978 and is projected to increase by 50 percent more in real terms by 1984.

### International Legitimacy and Imprimatur

International organizations can, and do, influence international relations (and national policies) in food and agriculture by legitimating practices and patterns, thereby often turning them into norms. The decisions of international organizations tend, as a rule, to evoke deference (even when they are honored in the breach), and the imprimatur "multilateral" tends to heighten the acceptability of practices prescribed for some governments by others. Moreover, the legitimacy and force of prescriptions by international organizations are not limited only to treaties, conventions, and other formal acts of international law making, but to resolutions, expressions of multilateral consensus and even to views, recommendations, and espoused doctrines of high-level international civil servants.[26] Relevant examples of multilateral legitimation include UN and FAO efforts to shape "development" into an international responsibility, GATT's strengthening of free-trade principles in the international commercial system, and FAO's endorsement of the principles of the freedom of agronomic information and the transnational basis of agricultural research. It is not surprising that many principles and practices legitimated by international organizations are often simply multilateralized versions of policies and preferences of powerful member states, as was, for example, FAO's endorsement of U.S. positions on the legitimacy of surplus disposal via food aid.[27] Achieving such multilateralization is indicative of exercising leadership within organizations. In addition, there is something to be said for the importance of international organizations, and their capacities to legitimize, as states go about seeking foreign policy instruments. As preferences for diplomacy through international organizations become more pronounced, seeking multilateral imprimatur should become an increasingly important and efficacious foreign policy stratagem.

## U.S. INTERESTS AND WORLD FOOD ORGANIZATIONS

As reported earlier, U.S. support for international food organizations, most notably FAO, has declined in recent years. This has coincided with the increased activity of the poorest states in the United Nations, where various agencies charged with international welfare tasks have become primary arenas of debate between advocates of a New International Economic Order and their critics, notably including the United States. As many practices in global food trade and aid have become subject to increasing challenge, stress, and deviation--and as these have often been articulated vitriolically in organizational forums--political leaders in the United States (and other industrialized countries) have reacted defensively to protect their perceived stakes in the traditional status quo. Part of their strategy has been to deflate universal multilateral bodies and hence to dampen "populist" pressures by circumventing forums controlled by Third World majorities. Alternatively, U.S. spokesmen have sought to create (or revitalize) smaller specialized institutions with built-in veto opportunities, weighted voting, limited membership or limited authority, and to propose bilateral alternatives to multilateral programs, where "one-on-one" rather than "one-against-many" bargaining conditions would prevail. Simultaneously, the stances of the United States and allied industrialized countries in the larger multilateral forums have either hardened against reform or shifted into defensiveness, delay, and waffling.[28]

To the extent that such a retreat from leadership in food organizations has characterized United States behavior in recent years, intentionally or not, it should be reversed. Two lines of reasoning support this conclusion.

First, <u>the current directions and policies of major food organizations complement U.S. policies and further U.S. goals</u>. If major statements of U.S. foreign agricultural policy--and declarations concerning our overseas development policies more generally--are reliable descriptions of our goals, then it must be recognized that many world food organizations are targeted in directions we clearly prefer. For example, calls for more emphasis upon self-sufficiency through rural modernization in the Third World put forth by U.S. officials and endorsed by the World Food Conference are currently reflected in the enhanced development activities of FAO, in the newly kindled rural development interests of the World Bank Group, in the World Food Council's prevailing concerns with "food security," in the plans and early allocations of the International Fund for Agricultural Development, and in the expanded resources and efforts of the World Food Program. Relatedly, U.S. interest and emphasis upon rural reform in the interest of poor farmers in poor countries, as reflected currently in Title III provisions in PL 480 and more traditionally contained in guidelines for the Alliance for Progress and other U.S. assistance programs, are at the center of FAO concern and World Bank, WFP, and IFAD planning.

In a similar manner, U.S. concern for enhanced world food production through agronomic research and technological innovation is matched by endorsements for research affirmed at the World Food Conference, and has strengthened programs and funding via CGIAR. President Carter's election campaign called for making the United States the breadbasket, not the arms merchant, of the world. This was not only . . . followed up in specifics by the United States, as in Washington's more aggressive efforts to get larger food aid commitments, or in the 6 million ton emergency reserve before Congress since 1977, but these . . . also found counterpart articulations in the deliberations of the World Food Council and in multilateral negotiations surrounding the renewal of the International Wheat Convention. They were also contained historically in FAO's design for a World Food Authority . . . in that organization's 1975 endorsement of an "International Undertaking on World Food Security."[29]

With regard to the international commercial system for food, current GATT concerns, reflected in the Multilateral Trade Negotiations (MTN) agenda, about abusive marketing practices and nontariff barriers to agricultural trade accord well with U.S. interests.[30] FAO studies on Agricultural Adjustment have tried to pinpoint areas where trade barriers might possibly be reduced.[31] Additionally, FAO's creation of an "early warning" mechanism and steps taken in Rome to enrich and expedite reporting on crop and marketing conditions also complement U.S. interests and needs.

U.S. concerns for humanitarian food distribution and the nutritional upgrading of diets among the world's poor are promoted by the World Food Program, the World Health Organization, the United Nations Disaster Relief Organization, and other organizations currently moving to mobilize and systematize food aid globally.

To indicate convergences and congruences between U.S. goals and the programs of international organizations is perhaps both obvious and also beside the point. Many of the international programs noted are in fact the past fruits of U.S. leadership, and their relationship to U.S. preference is therefore hardly surprising. Moreover, it is not so much these programs that are the sources of present U.S. dismay with the organizations, but rather pending and accomplished changes in the authority and procedures of the institutions themselves. High politicization and ideological conflict characteristic of North-South relations in recent years have penetrated the food institutions, and the developing countries' clamoring for "participatory democracy" has indeed fouled multilateral decision making. The United States' frustrations in this new international political setting are real. Still, is opting out of a game that looks so promising in the potential outcomes preferable to learning patiently how to play it better? If U.S. interests really rested in maintaining a concentration of decisional influence in international organizations that is unreflective of the distribution of global influence, or in defending a status quo biased against the Third World, or in instituting an empire founded on "food power," then withdrawal toward unilateralism might be in order. But, if these latter are not U.S. interests, there is little to be lost, and a great deal to be gained with regard to our substantive interests in global food trade and security, by adjusting tactics and positions in institutional forums and better supporting promising multilateral programs.

Second, international organizations provide promising arenas and instruments for the pursuit of U.S. food goals. Another strand in reasoning for positive U.S. participation in international food organizations concerns limitations on the United States' capacities to accomplish ends unilaterally. It is not necessarily the case that U.S. power in world affairs has declined, at least not in world food affairs. Indeed, this country continues to command both the resources and the allocative flexibility to support an active and effective foreign policy, and in the food sector our resources are even more abundant than during the heyday of ascendance during the 1950s and 1960s. Moreover, our vulnerabilities in global food affairs that stem from dependence upon commercial markets are likely to remain more theoretical than real as we move into the predicted high-demand decade of the 1980s. Ideally, then, the United States commands most of the ingredients for a unilateral foreign agricultural policy motored by capacities to reward partners who support our goals and to penalize detractors, much as was done during the 1950s and 1960s.

But the ingredient missing is legitimacy. The success of U.S. policies, including our food policies, in the decades after World War II followed in part from our command of and ability to allocate resources, that is, from our power, so to speak. But it also followed in part from widespread international acceptance of the legitimacy of our pursuits and equally widespread perceptions that U.S. goals complemented global well-being. Needless to say, these acceptances and perceptions have weakened in the face of Third World suspicions and insecurities, assaults from ideological antagonists, debacles in U.S. behavior, and manifestly diminished self-esteem. The United States can no longer expect others to support our policies or acquiesce in our goals because of their self-evident legitimacy, and we can no longer necessarily expect others' compliance and cooperation in deference to our respectability. Because of this, the effectiveness of U.S. unilateralism has been blunted.

Because international organizations are the primary legitimating agents in contemporary international relations, especially in Third World eyes, their multilateral endorsements are crucially important to the success of U.S. foreign policy, including food policy. In effect, it is increasingly important that the resolutions, decisions, and recommendations of international organizations coincide with U.S. ideals and preferences. This is not to say that the United States must defer to the whims of international parliamentary majorities. Nor, most emphatically, does it mean that the United States should finance unpromising international schemes or assimilate values alien to our traditions. Quite to the contrary, bringing greater coincidence between U.S. preferences and international organizational outcomes means that U.S. officials must more effectively influence international institutional processes, through more active, more positive, more creative, and, if necessary, more aggressive and dogged participation. Elements of U.S. power remain available--resources, money, and expertise. The challenge in international food affairs for the 1980s is the need to learn how to use these effectively within international organizations to foster cooperative action in preferred directions reinforced by global legitimacy.

## U.S. POLICY AND INTERNATIONAL ORGANIZATIONS: RECOMMENDATIONS FOR ACTION

As noted, in many areas, achieving coincidence between U.S. international food objectives and international organization programs means keeping the world institutions running on tracks already laid by constructive U.S. leadership during the last two decades. That is, it is in the U.S. interest, and it should be an object of U.S. efforts within international organizations, to bring more support for rural development in the Third World, more attention to small farmers and their needs, more concern with nutrition as a public policy problem, more food security for deficit countries, more research and information concerning agronomic and marketing predictions and problems, more stability in the international market, and more attention to distortions and abuses of free trade.

The United States should also make efforts to improve the operating efficiency and effectiveness of world food organizations. First, there has been great proliferation in the number of food organizations and programs in recent years. This is the result partly of frustrations at the slow pace and apparent ineffectiveness of some of the older institutions, partly the result of enthusiasm generated at the World Food Conference, and largely the result of various coalitions' attempts to structure arenas that are politically and procedurally biased in preferred directions. Obviously, large, industrialized countries feel more comfortable in small, limited-membership organizations with either weighted voting or consensual decision making, while small Third World countries prefer large parliamentary bodies that decide issues on the basis of majority voting. But the result of institutional proliferation, as noted earlier, has been redundancy and confusion in assignments (for example, at least five major organizations are currently debating the issue of food reserves; nine are concerned with food aid; more than a dozen are involved in rural modernization). But, even more seriously, proliferation has brought about a dispersal of resources, as well as considerable waste on redundant administrative and overhead facilities. Greater efficiency in the use of development, food aid, and research funds could be achieved by streamlining the system of world food organizations and better specifying mandates. (This does not imply proliferating further by creating more umbrella organizations or coordinating boards.) While there are obvious political difficulties for the United States in pursuing world food goals through FAO, it is also the case that this organization is structured and mandated to carry out almost all of the tasks that we might like to assign to international organizations. FAO, moreover, has traditional stature and carries considerable respectability in the eyes of many of its members, the Third World states, in particular. Therefore, it is to be recommended that the United States make efforts to strengthen FAO, to concentrate international programs in food and agriculture within FAO, to encourage membership on the part of the Soviet Union, and to rise again to creative leadership within this organization.

Second, the United States should encourage international food organizations to engage in more frequent, more thorough, and more critical evaluations of their programs, with a view toward improving promising ones and eliminating unpromising ones.[32] For institutional and political reasons, reliable self-evaluation is one of the most difficult tasks for any large bureaucratic organization to accomplish. Biases toward self-praise, circumvention, statistical manipulation and inundation, and political sterilization are pronounced in organizational reporting, and the frequent result is that major problems accumulate unattended until crises occur and resources are wasted. In addition, since the funding and functioning of many world food organizations, particularly FAO, depend upon assessments and subscriptions that must be periodically approved by national governments, continuing support depends upon demonstrations of performance in the form of rich, accurate, and freely flowing information about programs and activities. The U.S. government, and others as well, should be firm in demanding strict accountability from world food organizations, not in the interests of constraining them, but rather toward the end of strengthening them.

U.S. leadership and effectiveness within international organizations depend in considerable measure upon the reasonableness and coherence of our national positions. Yet, there is constant criticism, both within Washington and from other governments and international officials, that U.S. stances tend frequently to be internally inconsistent, diffuse, ambiguous, and varied from organization to organization.[33] The manifest split in the U.S. delegation to the World Food Conference, for example, is a dramatic and embarrassing case in point. Other examples of apparent inconsistency include the U.S. emphasis on the efficacy of the "free market," frequently underlined in our international organizational stances, set beside our willingness to negotiate multiyear bilateral grain agreements outside of the market, or, U.S. reservations about "commodity agreements," expressed at UNCTAD, set beside our efforts at the International Wheat Conference to create managed food-grain reserves. There may be some underlying consistency in such behavior, but its outward incoherence is understandably puzzling to those with whom we deal. . . . Greater efforts must be made to integrate the positions of different U.S. executive agencies that address food issues, and to better coordinate the articulation and the funding of food policies in relations between the executive branch and Congress. We do not underestimate the difficulties in such integration, since the United States is not easily constrained to speak with only one voice in either domestic or foreign affairs. Yet, our capacity to lead in international organizations is unquestionably related to our ability to point in clear, unambiguous directions.

Finally, there is the question of the number and quality of people that the United States contributes to the international civil service. Among criticisms of U.S. effectiveness in FAO, for example, are complaints that there are too few U.S. representatives in the FAO secretariat, that some of our nationals working there are distinctly less competent than might be desired, that incentives for younger U.S. officers to enter the international civil service are absent or comparatively unappealing, and that senior technical experts of high competence cannot be recruited for overseas assignments because alternative opportunities in the United States are invariably more attractive to them.[34] The result of this combination of factors is declining U.S. professional influence within international food organizations and increasing dominance of leadership positions by non-Americans. In light of this, and in view of our recommendation that the United States give greater weight to international organizations in its food diplomacy, we urge: (1) that the U.S. government move rapidly to fill currently vacant positions allotted to our nationals; (2) that it move to expand opportunities for Americans in international civil services, at least to degrees commensurate with our budgetary contributions; (3) that it devise incentives for young officers to improve comparative attractiveness of international as opposed to national service; (4) that it support the upward mobility of competent Americans within international civil services; and (5) that it improve the attractiveness of international assignments for senior experts by providing income or other incentives. Competent people have traditionally been one of the United States' most valuable foreign policy resources. Every effort should be made to make optimum use of these resources in the interest of international organizational effectiveness, and in the service of U.S. international food policies.

## NOTES

1. U.S. Senate, Select Committee on Nutrition and Human Needs, <u>The United States, FAO, and World Food Politics: U.S. Relations with An International Food Organization</u> (Washington, D.C.: Government Printing Office, June 1976), pp. 11-13. Hereafter this Senate document will be cited as the <u>Percy Report</u>.

2. Martin Kriesberg, <u>International Organizations and Agricultural Development</u> Foreign Agricultural Economic Report No. 133 (Washington, D.C.: U.S. Dept. of Agriculture, Economic Research Service, May 1977), pp. 31-33.

3. Ibid., p. 36.

4. Ibid., p. 32.

5. World Food Conference, <u>Report of the World Food Conference</u> (Rome: 1974), E/Conf. 65/20, p. 18; See also Robin Sharp, "The World Food Council: More Than a

Garnish?" Food Policy 1 (Nov. 1975):84-86; Thomas G. Weiss and Robert S. Jordan, The World Food Conference and Global Problem Solving (New York: Praeger, 1976), p. 68.

6. Report of the World Food Council on Its Fourth Session, June 12-15, 1978, General Assembly Official Records, 33d Session, Supplement #19, A/33/19.

7. World Food Conference, Report of the World Food Conference, p. 13.

8. Consultative Group on International Agricultural Research, CGIAR (New York: CGIAR, 1976), pp. 4ff.

9. B. Gosovic, UNCTAD, Conflict and Compromise: The Third World's Quest for an Equitable World Order Through the United Nations (Leiden: Sijthoff, 1975), pp. 93-114 and passim.; J. C. Nagle, Agricultural Trade Policies (Lexington, Mass.: Heath, 1976), pp. 70-79.

10. Percy Report, pp. 25-28.

11. Several United States delegates to FAO meetings interviewed in the course of our research, especially individuals who attended during earlier periods, expressed dismay and discomfort at the aggressive and "unsound" views put forward by many of the newer nations' representatives.

12. Hearings before the Senate Select Committee on Nutrition and Human Needs on U.S. Participation in the Food and Agriculture Organization of the United Nations, 94th Cong., 2d sess., March 4-5, 1976, pp. 93-94.

13. Ibid., p. 94.

14. Ibid., p. 105.

15. Ibid., pp. 118-19.

16. FAO, Proposal by the Director-General on International Action to Assure Basic Food Stocks, FAO Council, 60th Session, CL60/11 (Rome: Food and Agriculture Organization May 1973); World Food Security, Proposal of the Director-General, FAO Conference, 17th Session, C73/17 (Rome: August 1973).

17. OECD, Agricultural Policy in the United States (Paris: Organization for Economic Cooperation and Development, 1974), p. 103 and passim; OECD, Agricultural Policy in Canada (Paris: OECD, 1974), passim.

18. I. M. Destler, "U.S. Food Policy 1972-1976: Reconciling Domestic and International Objectives" International Organization 32 (Summer 1978):617-44.

19. U.S. Congress, Office of Technology Assessment, *Food Information Systems* (Washington, D.C.: GPO, June 1975), pp. 61-70; See also, *Food Information Systems*, Hearings before the Technology Assessment Board of the Office of Technology Assessment, 94th Cong. 1st and 2d Sessions, September 24, 25 and December 10, 1975, February 4, 1976 (Washington D.C.: GPO, 1976), pp. 70-81 and passim; David A. Kay, "The Behavior of International Organizations in the World Food Arena," in *The Management of Food Policy* ed. Giulio Pontecorvo (New York: Arno Press, 1976), pp. 123-25.

20. Joseph M. Jones, *The United Nations at Work: Developing Land, Forests, Oceans and People* (Oxford, England: Pergamon Press, 1965), p. 118ff.

21. Consultative Group, GGIAR, pp. 11-67.

22. Lester Brown, *Seeds of Change* (New York: Overseas Development Council, 1970).

23. Kriesberg, *International Organizations*, passim.; Jones, *United Nations at Work*, passim.; USDA, *Multilateral Assistance for Agricultural Development* (Washington, D.C.: U.S. Department of Agriculture/ERS, October 1973).

24. *Percy Report*, pp. 20-25.

25. Kriesberg, *International Organizations*, p. 68; Montague Yudelman, "The World Bank and Agricultural Development," in *Management of Food Policy*, ed. Pontecorvo, pp. 17-36; Susan George, *How the Other Half Dies* (Harmondsworth, Middlesex, England: Penguin Books, 1976), pp. 235-64.

26. Inis Claude, *The Changing United Nations* (New York: Random House, 1967), pp. 74-104.

27. Robert L. Bard, *Food Aid and International Agricultural Trade* (Lexington, Mass.: Heath, 1972), pp. 111-30.

28. *Percy Report*, pp. 3-4.

29. Helge Ole Bergesen, *When Interdependence Doesn't Work: A Study in World Food Politics* (Oslo: Norsk Utenrikspolitisk Institutt, 1977), pp. 89-93ff.

30. See Chapter 2 above; Simon Harris, *E.E.C. Trade Negotiations with the U.S.A. in Agricultural Products* (Ashford, Kent, England: Centre for European Agricultural Studies, 1977), pp. 9-14.

31. See Chapter 2 above; Bergesen, *When Interdependence Doesn't Work*, pp. 158-61.

32. U.S. Participation, Hearings, pp. 78-86.

33. Ibid., pp. 1, 16, 127ff.; Percy Report, pp. 2-3.

34. U.S. Participation, Hearings, pp. 78-86 and 107.

# 4

# SOME ISSUES IN THE MULTILATERAL MANAGEMENT OF FOOD RESOURCES

*Janet K. Schmidt*

## THE DEVELOPMENT OF U.S. POLICY TOWARD MULTILATERAL MANAGEMENT OF FOOD RESOURCES

Ever since the end of the Great Depression, the United States has been the most important producer and trader of food in the world. Almost half the grain in the world trade comes from the United States and at least 80 percent of the soybeans. During some periods the United States has provided as much as 90 percent of the world's food aid. About one-third of the U.S. agricultural product is sold for exports. In 1982 those exports contributed almost $44 million to the U.S. balance of payments. Nevertheless, the abundance of the U.S. agricultural products has sometimes seemed like more of a curse than a blessing.

### The Problem for the U.S. Farmer

The abundance of potential supply has made the U.S. farmer susceptible to erratic cycles of boom and bust. When weather and other conditions are favorable, production greatly exceeds effective demand and the market price of the commodity falls, sometimes even below the cost of production. The most vulnerable farmers go out of business, and the remaining farmers are faced with only risky options.

Every year each of these farmers must decide how much to plant. If all farmers cut back production, prices rise. But if only <u>some</u> farmers cut back production and others do not, those who do not could be driven out of business. The farmers who plant from fence to fence could reap huge profits, at the expense of those who cut production, and expand their businesses the next year. But if these farmers have gambled wrongly and production is high elsewhere too, prices will head downward, perhaps to a point below their "breakeven point." Since they have invested more, they lose more.

In other words, both very large and very small crops can cause economic disaster. Very large crops drive prices so low that the crop cannot be sold at a profit. If crops are large for several years, the number of farms and farmers could be so reduced that when a cycle of bad growing conditions occurs, the remaining farms would not be able to produce enough food. Very small crops have equally dire consequences. The elasticity of demand for food is such that a decline in supply can push prices up very high very rapidly as they did in 1973 when commodity prices tripled in a few months. Under such conditions the farmers have every incentive to invest in larger crops in subsequent years. But they do not have the information they need to make rational market decisions. They do not know what the aggregate effect of other farmers' decisions will be. They do not know what the growing conditions will be. And they do not know what other governments' policies will be. They may help turn a boom into a bust or a bust into a boom, but they will not know until the end of the growing season.

In any case, farmers are not the only people affected by the wide swings in the fortunes of the agricultural sector.

## The Problem for the Economy

What is for the farmers a set of conditions about which they cannot calculate risk very accurately is for the country a serious problem for the economy as a whole. The agricultural sector of the U.S. economy is a much more influential sector than most urbanites suppose. It accounts for about 20 percent of the U.S. gross national product and more than 20 million jobs. In recent years the sale of U.S. agricultural products grossed $45 billion and accounted for 20 percent of all U.S. exports.[1] The wide price fluctuations for agricultural products cause problems throughout the rest of the economy. High food costs inflate the cost of living and eventually the cost of labor throughout all sectors of the economy. High labor costs are passed on to the consumer and, since farmers are consumers of more expensive agricultural inputs, the cost of planting and harvesting the next crop is greater. Thus high agricultural product prices are highly inflationary.

Any country in which agriculture is a significant proportion of the gross national product will experience the random and ripple effects of agriculture's market swings. U.S. agriculture has seen a long series of booms and busts, but no agricultural crisis prompted large-scale government intervention until the 1930s. The crisis of the 1930s, however, was so severe that it threatened irreversible damage to both the economic and ecological viability of the industry. Agriculture was too volatile to attract new investment. The existing investment in the land was being lost as one farmer after another overtaxed and then abandoned the land to the dust storms.

Eventually the U.S. government took the position that no great country could allow its food supply, and consequently its economy, to be put in jeopardy because of the volatility of agriculture. Subsequent administrations faced the fact of market failure and sponsored legislation to regulate the production and marketing of agricultural commodities. Extension services grew and Americans began to take for granted that it was a legitimate prerogative of government to concern itself with the health of this prerequisite of all industries.

During World War II the U.S. government did not need to do very much to help agriculture. The worldwide wartime demand for agricultural products was enormous, and the capital accumulating in U.S. agriculture was invested in boosting production. Once the security crisis was over, however, foreign governments could no longer justify U.S. control over their own agricultural markets. Other countries began protecting their agricultural sectors from cheap U.S. competition. U.S. farmers recognized that the world market was gradually shrinking, and they modernized for greater efficiency.

By the late 1950s the U.S. agricultural surplus was being stored in every conceivable receptacle, including decommissioned warships, and was driving down the value of new production. Even the most efficient, modern farmers were finding it difficult to stay in business. The Eisenhower Administration was forced to find a way to support farm prices and reduce surpluses. Nevertheless, the irony of spending millions of tax dollars to discourage the production of food, or to store it indefinitely while millions of people in the world were undernourished, was not lost on the U.S. government. It devised a solution to the domestic political and economic problem posed by the agricultural sector's volatility.

In so doing, the government thrust the United States into interdependence with all other important producers and consumers in the world agricultural market. Policies devised initially for the improvement of the agricultural sector of the U.S. economy would have quite a profound influence on the agriculture policies of U.S. trade partners and aid recipients. Other countries responded to U.S. policies with trade, aid, and agriculture policies of their own.

## The Problems of Trade, Aid, and Foreign Influence

Two legislative remedies for the U.S. food glut were designed to get food to the poor without spoiling the commercial market for food products. The first of these programs distributed food to the poor in the United States and was gradually transformed into the Food Stamp Program.

The second program came to be known as "P.L. 480" or Public Law 480. It made U.S. surpluses available to foreign governments on concessional terms. U.S. grain could be sold to a country on subsidized credit or donated to relief organizations that would then distribute the food to the needy. At present a little more than two-thirds of the P.L. 480 budget is allocated to fund Title I concessional sales. The remaining third is for Title II food donations. The current P.L. 480 budget is about $1.5 billion.

P.L. 480 sales and donations have made the United States the most important actor in international food negotiations. While the enormous U.S. surplus has given the United States considerable "leverage" in international negotiations (e.g., Egypt, Israel, Pakistan, India, Bangladesh), P.L. 480 has also made the smooth operation of U.S. surplus food disposal policy more dependent on other countries' agricultural policies. U.S. agriculture has become accustomed to huge government-subsidized foreign sales and gifts. Without them a major transformation of U.S. agriculture would be unavoidable.

The United States wanted to dispose of its surplus in the least expensive way. The obvious thing to do with it was to export it. But one of the reasons the surpluses existed was that there was not a market for the product at the price U.S. farmers needed to cover the costs of production. Exports had to be subsidized. Subsidizing exports helped the government to achieve its objective of boosting farm income, but it was incompatible with the government's goal of cutting spending. Farm aid rose abruptly, not in spite of huge exports, but, in part, because of exports. In 1974 farm aid amounted to $3 billion. In 1982, only eight years later, farm aid was four times 1974 levels ($12 billion). In 1983, the farm aid appropriation was $21.2 billion, almost twice the 1982 expenditure and more than seven times the 1974 appropriation.

The reason for this huge increase in farm aid is that "the price per bushel of the good sold in the export market is too low to cover the full production costs. For wheat, the price received from exports covers only 65 percent of the high cost of production."[2] The difference between production cost and market price is made up by the government, occasionally through explicit export subsidies but more frequently through subsidizing inputs and through price supports.

As the cost of exports increases, so does the domestic economic problem of paying for the increase. The long-standing domestic political commitment to support farm income is incompatible with cutting spending. But cutting spending on agricultural supports would have long- and short-term political and economic costs that are unacceptable. It would involve a costly restructuring of resources that would almost certainly (1) fuel inflation by substantially raising the cost of food to consumers, (2) decrease the number of farmers, and (3) decrease the demand for agricultural inputs.

If it were politically possible, letting the domestic farm economy "revert" to market conditions sounds like it might solve the budget deficit problem--but it would create other difficulties. The United States would have to be willing to let its huge trade deficit ($27.3 billion in 1981) grow. The problem would not be solved. The burden of its solution would just be shifted from the officials who are responsible for the budget to the ones who are responsible for trade balances.

In fact, it is essential that one grasp this aspect of the agriculture problem: That proposed solutions merely shift the burdens and sacrifices from one group to another; from farmers to taxpayers, from producers to consumers, from U.S. to foreign interests and back again.

## The Soviet Grain Deal and the Interrelatedness of Farm, Economic, and Foreign Policy Problems[3]

In 1972 the Nixon Administration believed that the Soviet Union would encourage the North Vietnamese to make peace with the United States if the United States made concessions to the Soviets. Kissinger believed that he could persuade the Soviets that there were more benefits to be had from cooperation with the United States than there were in sponsoring national liberation movements. In addition, Nixon had been looking for ways to discharge his debts to his Midwest constituency for its support in the 1968 elections. He proposed to help the U.S. agricultural sector by expanding export markets for U.S. grain. At the summit meeting between Nixon and Brezhnev in 1972, the United States agreed to allow the Soviet Union to purchase $750 million worth of grain over the next three years. I. M. Destler describes what happened this way:

> Soviet Exportkleb representatives had quietly contracted with the Continental Grain Company to supply them 4 million tons (147 million bushels) of corn. And they made this purchase not on credit but for cash. Information about this record sale leaked out gradually over the next month, during which the Soviets purchased an additional 7.8 million tons (286 million bushels) of wheat from Continental and five other grain companies. They also purchased about 2 million additional tons of feedgrains and 1 million tons of soybeans.
>
> These amounts were several times what anyone had expected. The 433 million bushel total wheat purchase, for example, amounted to half of U.S. carryover stocks of July 1, 1972, and a bit over one quarter of 1972 production. . . . By sometime in August evidence was overwhelming that the world wheat situation had been transformed from a buyers' to a sellers' market.[4]

The deal furthered the U.S. goal of developing new trade outlets, but it resulted in a tripling of grain prices in two years. It set off a round of inflation that forced senior economic advisors to realize that liberal export policies, coupled with production cutbacks in agriculture, could ruin any plan for economic stability. Between 1972 and 1980 food prices consistently increased faster than nonfood prices.

In addition, the slow draw-down of reserves in the United States caused a huge increase in the price poor countries would have to pay for their food imports. Drawing down U.S. reserves was designed to cut domestic expenditures on the storage and administration of the surpluses. Unfortunately, it coincided with bad weather in a number of other exporting countries. By 1974 world food reserves had been reduced to a mere 33 days.[5] The 1972-74 tripling of grain prices caused severe hardships in many poor countries. It is estimated that millions of people died of malnutrition during the crisis. It is impossible to calculate how many more died from the overall impoverishment of these countries in terms of development opportunities forgone because of the dire need to spend more on food imports. To make matters worse, the amount of food available for P.L. 480 distribution had always been determined by how much money was budgeted by Congress for P.L. 480. So when food prices tripled, the amount of food poor countries could buy or even receive in emergency aid was severely and automatically cut. "For example, $655 million bought about 9.1 million tons of P.L. 480 grain in FY 1971; $727 million bought only 3.4 million tons in 1974."[6] One-sixth of the volume of food stuffs cost three times as much in dollars.

Even before the full impact of the Soviet grain deal was felt, the food security of the traditional food importing countries had deteriorated badly. In the late 1960s food prices rose in the world market; importing countries adjusted. When Nixon devalued the dollar in 1971, they adjusted again. Then the OPEC nations announced their increase in the price of petroleum. By the time the Soviet grain deal tripled food prices, many poor countries were already much worse off than they had been in the early 1960s. Several countries suffered differing degrees of social unrest due to inadequate provisioning of food.[7]

Nevertheless, the U.S. Department of Agriculture under Earl Butz took the position that the department's problem, disposing of surpluses, had been solved. It proposed a USDA budget without any allocation at all for P.L. 480, not even for Title I concessional sales.[8] The budget allocations for P.L. 480 were subsequently revived. Nevertheless, in the last half of 1974, U.S. concessional sales under Title I amounted to only one-sixth of the metric tons delivered two years before.[9]

The Soviet grain deal illustrated to anyone who had not yet grasped the nature of the problem that farm, economic, and foreign policy problems were interrelated in such a way that attempts to solve any one of the problems without reference to any of the others was courting economic, political, and diplomatic disaster.

## The Limits of a Free Market Solution

Even if the economic stability of poor Third World countries really were not of any strategic interest to the United States, the same cannot be said of the European Community, Japan, China, India, and the Soviet Union.

The emergence of the European Economic Community's Common Agricultural Policy (CAP) over the last twenty years has gradually reduced Europe's dependence on U.S. agriculture. The United States complains, sometimes quite bitterly, about the Community's policies for protecting agriculture in Europe. Nevertheless, the countries of the EEC were more able to withstand the economic shock of the 1972-74 world food crisis because of CAP policy. All but the most hard-line free marketeers appreciate the point of view of the Community's leadership. The Community acts on what is in the United States a well-accepted principle: that the health of a country's agricultural sector is crucial to the vitality of the economy as a whole.

The Japanese import a higher percentage of their food than any other modern industrialized country (27 percent). Japan's own agricultural sector is small and would be destroyed without protection, leaving Japan dependent on foreign countries for its food supplies. Japan is an important ally. It is not in the interest of the United States that it be any more dependent on other countries for its domestic food supply.

It makes no sense at all to insist that the Soviet Union, China, and India adopt a free market strategy with respect to the allocation of scarce food resources. It is precisely because these countries believe that a free market in food will not serve their interests nationally or internationally that they have adopted nonmarket strategies. If one were to subtract Europe, Japan, the Soviet Union, China, and India from the world market, there would be very little agricultural trade left to be "free."

Under the circumstances, the U.S. claim that free markets would serve everyone's interests far better than protectionist policies sounds both unrealistic and disingenuous.

The free market strategy does not work either domestically or internationally because of the special character of the agricultural sector. The market for agricultural goods has three features that preclude its performing the coordinating and rationalizing functions that are necessary to stabilize this volatile sector of the economy. First, the weather so substantially affects supply each year that it is impossible to predict the market price of commodities at the end of the growing season. Second, other governments' growing, buying, and selling policies can change dramatically and without warning. Some countries even treat the size of their crop as a state secret, making it almost impossible to estimate world demand. Third, the market failure so characteristic of the agricultural sector has even more unfortunate effects on the economy as a whole because of inelasticity of demand for food. Many other industries are affected by the supply and demand for agricultural products. Industries that produce fertilizers, farm machinery, and seeds or that process and market commodities all require some stability in agriculture to be healthy industries themselves.

In fact, conditions endemic to agriculture make it impossible to have a stable, free market. Agriculture exhibits all the classical symptoms of market failure; imperfect information, the collective goods dynamic, monopolistic or oligopolistic control, and resource immobility. Supply and demand do not rise and fall freely. Resources are not shifted so as to increase the economic welfare of some without causing harm to anyone. In other words, not enough prerequisites for a free market exist to produce the proposed benefit of a free market--efficient allocation of resources.

Advocates of a free market might argue that no market is absolutely free, that "free market" is merely an ideal type, that real markets only approximate the efficiency of a free market. But such an argument avoids the issue, which is that none of the principal actors is willing to leave such important issues as the economic health of the agricultural sector or food security to the "invisible hand." The European Community, Japan, the Soviet Union, Canada, Australia, India, China, and even the United States have elaborate and costly policies precisely because free markets have been too volatile. In fact, it is difficult to see how any of the goals of U.S. food resource management policy could be better achieved if there were a free market for agricultural commodities.

It was the Nixon Administration's attempt to increase U.S. agricultural exports that exposed the real dangers inherent in the market: world reserves dropped to a record low of thirty-three days and grain prices tripled. A politically unacceptable level of food price inflation was reached at home, while food aid and relief abroad was insufficient to substantially reduce widespread hunger and starvation. The farm economy improved but the improvement was short-lived; in 1983, just ten years after the Soviet grain deal, the farm aid appropriation was $21.2 billion, more than seven times the 1974 level. Nevertheless, the Reagan Administration's position has been that the market should be the rationalizing mechanism for agriculture. In fact, the current official U.S. policy is to encourage a free market in agricultural products both internationally and domestically. As one Reagan Administration spokesman said, the United States is "committed to building a liberalized world trading system which provides increased opportunities for U.S. agricultural exports."[10]

Up to this point we have examined the U.S. preference for leaving the rationalization of agriculture to the free play of market forces. This is not the whole story. The United

States has also been a major actor in organized, multilateral attempts to cope with the problems agriculture poses. For that part of the story we must return to where we began, the Great Depression and its impact on multilateral management of food resources.

## MULTILATERAL ATTEMPTS AT MANAGING FOOD RESOURCES

### The Founding of the Food and Agriculture Organization (FAO): The Mandate

The origin of the FAO is said to be a paper called the "McDougall memorandum" published in January of 1943.[11] FAO's biographer, Gove Hambidge, says that it was when this paper was shown to Eleanor Roosevelt that the idea of an international organization for food and agriculture made its first contact with the kinds of people who could actually bring such an organization into being.[12] For this reason the content and background of this paper are a very important clue to understanding what the founders of FAO thought they were doing. Hambidge reminds his readers that the Depression had thrown agricultural markets into chaos and that

> There were a few conflicting views and interests, mainly in the realm of economics. Some food-importing countries, for example, were highly dubious about the possibility of international action to protect producers if unmarketable surpluses should develop, because this might tend to raise prices. Some big production countries, on the other hand, were not cordially disposed toward the possibility of establishing international food reserves (buffer stocks) to protect consumers against shortages, because the practice might tend to depress prices.[13]

Governments wanted to prevent a recurrence of the prewar collapse of agriculture but were skeptical of concerted international action. Nevertheless, forty-five countries accepted President Roosevelt's invitation to a conference on food and agriculture held in Hot Springs, Virginia, in May of 1943. Hambidge says that despite many governments' doubts,

> The delegates nevertheless found more areas of agreement than disagreement. The fact that so many were technical people--experts in nutrition, in agricultural sciences, in fisheries, in statistics--rather than diplomats or politicians perhaps helped to make this agreement possible; for in the language of science, hunger and its effects are the same everywhere, and so is animal disease,

> and so is the breeding of better varieties of plants.[14]

It is significant that the author of this history of FAO attributes the success of the conference to the fact that the delegates were "technical people" because, as we shall see, this became the dominant theme of the organization.

It was in this atmosphere of despair over World War I, the Depression, the rise of fascism and World War II that the McDougall memorandum was written and read.[15] It was an era characterized by a profound disillusionment with the idealism of the interwar period. The faith in international deliberative institutions such as the League of Nations was shaken.[16]

In such an atmosphere, those who favored the reestablishment of international organizations such as the League had to make a different kind of argument than was made in the interwar period. They argued that the war was caused by economic chaos, poverty, and despair. In other words, they hypothesized a link between prosperity and peace. Stable, prosperous societies would avoid war; unstable, poor states would have little to lose in war.[17] Thus solving economic problems could prevent war. Politics and diplomacy were perceived to have failed. The alternative to international bargaining was "functionalism." Virtually every influential leader of a postwar international organization was a functionalist including FAO's first Director-General, John Boyd-Orr.[18]

> Its [functionalism's] fundamental premise is that nation-states are not the most promising instrumentalities for achieving an enduring peace, nor are conflict-laden issues of high politics the most promising vehicles for developing habits of cooperation. Eschewing states and statesmen, the functionalist emphasizes an incremental approach to peace which would be based on the collaborative efforts of experts in relatively noncontroversial areas of shared interest and need. By a process of spillover, the penchant for cooperation developed in one sector would lead to cooperation in other sectors. In the long run, states would atrophy and a nexus of cooperative relationships based upon function would be the characteristic feature of the international system.
>
> This millennial view of international relations patently overstates the capacity of experts, in the absence of benevolently active governmental support, to transform the world.[19]

Nevertheless, the founders of FAO believed they had found an alternative to politicians and diplomats in the "technical people."

An international civil service of experts would create the habit of cooperation among nations by demonstrating that conflicts of interest were merely "apparent." The experts would lead the way to international cooperation by gradually assuming authority over more controversial issues.

In the first annual publication of the Work of FAO, peace and freedom from want are declared interdependent.[20] And it would be the technical people who would be able to "lift the burden of hopeless poverty from men's shoulders by applying modern scientific knowledge to the development of the earth's resources."[21] FAO acknowledges that different states have different economic and political interests, but the first annual Work of FAO tells us:

> If there is any one fundamental principle on which FAO is based, it is that the welfare of producers and the welfare of consumers are in the final analysis identical . . . [T]he identity of producer welfare and consumer welfare is by no means always self-evident. On the contrary, the interests of the two often apparently conflict.
>
> But the conflict is only apparent . . . . Wherever the contrary seems to be true, it is because all of the factors have not been taken into account, including the risk of social upheavals and wars.
>
> There is always a larger framework in which producer and consumer interests are seen to be the same.
>
> It will be the business of FAO to seek and to emphasize this larger framework, this whole view.[22]

The larger framework, within which apparently conflicting interests can be seen to be identical, is the experts' technical framework. It helps to make agreement possible because it reveals what is universally true about such things as malnutrition, animal disease, and plant breeding. The experts, because they speak the "language of science," are thought to be capable of circumventing political and social conflicts and reaching agreement with experts from other countries. The experts understand the problem from the perspective of the "larger framework" wherein "interests are seen to be the same." The "language of science" gives experts an understanding of the problem that is not distorted by merely apparent conflicts of interest. Thus, if the experts are charged with improving the poor living conditions that lead to war, and if they do so by ignoring the social and political demands that arise from the misperception that interests conflict, then eventually people will come to see that the experts are better suited to defining the common interest than politicians and diplomats are. Eventually, the conflict over political and social interests will disappear because the problem of malnutrition will have been solved technically by the experts.

The founders are reasonable enough when they say that the social upheaval of the interwar period was related to the economic disaster in agriculture, but it does not follow that these evils arose from <u>ignorance</u> of the underlying common interest that might have been revealed by expert collaboration.

Nevertheless, McDougall, in his 1943 memorandum, and John Boyd-Orr, in his first speech to FAO after having been elected FAO's first Director-General, went even further and declared that both the nutrition problem and the economic problem could be solved together. McDougall quotes another functionalist, S. M. Bruce, an early and influential advocate of international cooperation on nutrition programs.

> Is it not possible that our newly acquired knowledge of the importance of nutrition to public health might open the way to a solution of the world's agricultural problem? Is it not the case that, today, most countries are facing a most paradoxical situation? Ministers of Agriculture and their advisers are wrestling with the problems of apparent overproduction. Increased yields are being regretted and abundance is . . . often officially deplored. At the same time, Ministers of Health and their official and medical advisers are realizing, more and more, that public health demands an increased consumption of many of the very products about which the Departments of Agriculture are so unhappy.
>
> Millions of pounds are being spent annually in subsidies, bonuses, and other forms of assistance to agriculture. Side by side with that expenditure millions of pounds are being devoted to combatting disease. Is it not possible to marry health and agriculture and, by so doing, make a great step in the improvement of national health and, at the same time, an appreciable contribution to the solution of the agriculture problem?[23]

McDougall himself says, "The adoption of sound nutritional standards on a worldwide basis would have a highly favorable effect on world agriculture and world trade."[24] The connection being made here between solving the nutrition problem and the world economic problem of agriculture was sufficiently important to be included in the preamble to FAO's constitution. The preamble says that the task of FAO is "bettering the conditions of rural populations and thus contributing toward an expanding world economy."[25] But FAO would not be given the power to coordinate international trade in agricultural commodities. "The Bruce-Orr-McDougall ideology was eloquently expressed in the preamble of the FAO Constitution, but the delegates were not in sufficient agreement to include it in the operative portions of the document."[26]

> Lord Orr had been identified with a group of Australian agricultural specialists before and

> during World War II, most prominently Lord Stanley Bruce and F. L. McDougall. These men were eager to bring about 'the marriage of health and agriculture' by creating an international structure that would not only dispose of agricultural surpluses, but combat malnutrition at the same time. The same idea was pressed by Orr after he became head of the new FAO: he proposed the creation of a World Food Board that would stabilize prices by buying up surpluses in time of glut, and would dispose of them, subject to health and nutritional standards, in time of famine . . . . Yet the FAO turned down the Orr Plan, and the Director-General resigned. Why? . . .
>
> Orr had plenty of warning at the meeting at which the FAO was set up that the major agricultural exporting nations were not ready to consider an international commodity policy; but he did not heed it.[27]

McDougall, Boyd-Orr, and Bruce were not ignorant of the interests involved; they just believed that they had discovered a solution to the problem that revealed a common interest underlying the "apparent" conflict. The founders did not deny that perceived conflicts of interests were a formidable obstacle to cooperation, but they believed that this obstacle was only apparent, that it only existed because of a misunderstanding or a too narrow understanding of the nations' real interests. There is the unmistakable tone that more knowledge and education will reveal common interests where merely "apparent" conflicting interests have obscured the solutions. Bruce, Boyd-Orr, and McDougall were not able to convince the states that FAO should be given the power to buy and sell commodities, but their arguments did persuade the states to include the functionalists' goal in the preamble. Their goal was the establishment of nonpolitical institutions. McDougall says:

> There will be many difficulties to overcome before freedom from want of food is a realized ideal. These include the innate conservatism of farmers and peasants, the opposition of vested interests, and, perhaps most formidable of all, the unawareness of certain types of politicians and bureaucrats of the need for change.[28]

If we apply the reasoning used in the Work of FAO quoted earlier, conservatism of farmers and opposition of vested interests are due to the failure of farmers and interest groups to take "all of the factors into account" and to adopt the "larger framework in which producer and consumer interests are seen to be the same." "Unawareness" in this context seems to suggest that knowledge and education are needed. The difficulty that must be overcome is the ignorance of the "need for change."

Hambidge tells us that there were two points of view expressed at the Hot Springs Conference about the nature of the organization. The first group, including Boyd-Orr, favored a "strong food and agriculture organization which could take positive steps to foster economic expansion and help prevent disastrous crises, the recent depression being still much on men's minds."[29] However, Hambidge says, there was a second group that "wanted a rather narrowly limited fact-gathering and advisory agency which would be carefully insulated from positive action."[30] The second group won. FAO was established with such limited functions because countries were afraid that a more active mandate would lead the organization to actions that would prejudice the interests of the states.[31] The organization was thus established with

> three broad functions: to collect, compile, and interpret statistics relating to agriculture and food consumption; to organize a world service around the sciences affecting agriculture; and to provide technical guidance and help for the less developed countries by recruiting groups of agricultural experts who would be 'available to be sent to any part of the world to assist in improving agricultural methods, securing better land tenure systems, bringing about more effective marketing, or promoting agricultural education.'[32]

Even though Boyd-Orr, Bruce, and McDougall wanted an organization with the authority and power to execute the experts' programs, they were willing to settle for a fact-gathering and advisory agency because they believed that more knowledge was the key to unlocking the opposition of the "interests." The clear implication was that hunger was a technical problem, and as such, could be solved technically, without even raising political questions. Since malnutrition caused social and political conflict (in the same sense that bad living conditions cause war), and since nutritional science could be used to eradicate malnutrition (by discovering the proper amount and kind of food each person needed), and since this determination could be made objectively by experts, nations should allow experts to allocate the world's food production based on objective determinations of nutritional need.

Nation-states were "not the most promising instrumentalities for achieving enduring peace" because the perspective of the sovereign state was too narrow to see the real, underlying, common interest. Thus the formal solution to the problem that sovereignty presented was to limit the organization's activities to data gathering, clearinghouse, and technical guidance functions. These functions were declared "nonpolitical." Rules and procedures were established to confine the organization to the nonpolitical. Boyd-Orr wanted a "strong" organization that could take "positive steps," but he became the first Director-General of FAO, even though it did not have the mandate he wanted, because he believed that states would drop their objections to international action when they realized some of the benefits. He hoped that technical cooperation would eventually lead the way to political cooperation. Eventually, such cooperation would "make for cooperation in other spheres. It would be a step towards the evolution of the UN into a World Government, without which there is little hope of permanent peace."[33]

Indeed, this line of argument was so common at the time and was taken so far that one analyst claims that the first Director-General of the World Health Organization believed that "the expert in mental health can save the world from the obsolete politician."[34] FAO's functionalist founders admitted that it was not as simple as all that, but the fact that they made this kind of argument was important nonetheless because FAO was established as an information-gathering organization even though the mandate was to do something about malnutrition.

This belief, that technical cooperation would lead to political cooperation, formed the background of compromise between the states and the functionalist founders. The technical approach seemed to be a way around sovereignty that nations would accept. It was assumed that the "technical people" did know or could find out the answer to the difficult problem that starvation and agricultural glut presented, but since governments were not willing to let international organizations have decision-making authority, governments would simply let the technical organizations find the answers and then the countries would decide whether or not to implement them. The political and social causes of the problem were mentioned, but action on them by the agency was prohibited.

The United States and other big producers of commodities were able to resist efforts to give FAO the power to buy and sell commodities. But the idea that FAO should have a more active mandate to do something about hunger and poverty in the non-Western world lingered on. By the early 1950s the United Nations had passed a resolution establishing the Special Fund which enabled the specialized agencies to expand their prerogatives. FAO would have the power to act, though not on the issue of trade in commodities. FAO expanded from a "mere fact-finding body" to an agency mandated to give technical assistance. But the technical assistance work, known as the "Field Program," was still bound by a resolution restricting all agency action to the "truly international and nonpolitical."[35] The resolution states that:

> The technical assistance furnished shall (i) not be a means of foreign economic and political interference in the internal affairs of the country concerned and <u>shall not be accompanied by any considerations of a political nature</u>; (ii) be given only to or through governments; (iii) be designed to meet the needs of the country concerned;

(iv) be provided, as far as possible, in the form which that country desires; (v) be of high quality and technical competence.[36]

## FAO's Functionalist Legacy

First we saw that the source of inspiration for the creation of FAO was the founders' belief that difficult political differences could be circumvented if countries could be persuaded to choose the technical perspective as their own perspective. Then we saw that the founders believed that a major economic problem, that of overproduction, might be susceptible to a purely technical solution. We saw that the United States advocated the adoption of a functionalist mandate, but one in which the experts would only make recommendations, not binding policy decisions. Then we found that FAO was given responsibility to help "develop" the countries that requested assistance, but that this was to be accomplished without taking into account "considerations of a political nature." Thus, the most important thing to keep in mind about FAO's mandate is that political considerations, including the prerogative of sovereign states to define and pursue their own interests, were regarded as impediments to solving the problem of starvation in the midst of agriculture glut.

The functionalist and the free market approaches to food problems are similar in one important respect: Advocates of each approach assume that there is an objective perspective that states can adopt that will reveal the common interest of producers and consumers. If everyone adopted the free market approach, then the "invisible hand" would ensure that supply and demand fluctuate with respect to one another in such a way that a rational equilibrium would be maintained. The functionalist expert should be able to show that the adoption of the "larger framework," within which interests of both producer and consumer states are seen to be identical, would produce a set of straightforward and uncontestable solutions to the world food problem.

As we have seen, the United States has been the most important advocate of both the free market and the functionalist approaches to the multilateral management of food resources. The United States advocates these two approaches for much the same reason: they both promise to achieve some sort of rational management of food resources without any state having to give up sovereign autonomy to an international decision-making body. Both approaches appear to sidestep the conflicts among and within states over the allocation of food resources. Built into the free market and the functionalist approaches is the assumption that avoiding political conflict is not only possible but desirable. Both approaches hypothesize a nonpolitical solution toward which state behavior will tend once the irrational prejudices are removed. In its advocacy of a free market in food resources, the United States argues that the allocation of food resources can be rationalized when states remove the barriers to free trade. When the United States insisted that FAO's powers be limited to fact finding and technical assistance and that FAO be given no power to buy or sell commodities, it was expressing its view that the only objective and realistic approach to improving food resource management was to avoid questions of allocation. The United States had avoided all attempts to establish organizations with the authority to allocate food resources on the grounds that free market and functionalist approaches to food resource management were objectively better approaches than was political bargaining. Political bargaining would bring to the surface all the fundamental value differences, all the ongoing ideological disputes concerning the just distribution of food resources.

## The World Food Conference

By 1973 it was obvious that the United States was paying a very high price for its well-known reticence to coordinate allocation with other countries. The United States' response to the food crisis was Henry Kissinger's proposal that the United Nations sponsor a world food conference. In a confidential internal memorandum an FAO official analyzed the U.S. proposal.

> The most important positive feature of this proposal is that it has come from the United States in a major policy statement and the United States would therefore remain interested in the success of the Conference. In the past 25 years almost all important proposals for international food reserves--in some form or another or for a longer term international food aid policy--were thwarted because the United States has not been prepared to accept greater international control of its food and food aid policy.[37]

Of course, the United States was not prepared to substantially change its position on international food resource management, so the results of the World Food Conference reflected the U.S. restrictions. The principal aim of the new United States' initiative would be to rebuild food reserves by institutionalizing a system of nationally held reserves. Specifically, the United States proposed that a new system be established wherein participating countries would be assured access to nationally held commodities at the prevailing market price. Additional concessions would be made from members to countries in dire emergencies, but nonparticipants in the reserve system would be denied access to the reserves. An international emergency reserve system would be administered by the UN/FAO World Food Program.*

---

*The three main institutions created to facilitate the

The World Food Conference debate over the appropriate institutional response to the food crisis mirrored the 1940s debates over the degree of authority FAO should be given. The United States, along with the other Western industrialized countries, argued that a free market would provide the best long-term solution to the problem of managing food resources. In the 1940s and again in 1974 the West endorsed a free market allocation of resources coupled with an institutionalized clearinghouse for data gathering and the coordination of programs. FAO and the World Food Council would have no new prerogatives or powers to allocate or transfer resources. FAO and the WFCL would continue to study the problem, keep abreast of developments in different countries, make suggestions for improving the functioning and coordination of the aid effort, and, finally, fashion WFCL recommendations that all the countries would accept. Since WFCL effectiveness would be measured in terms of states' compliance with WFCL recommendations, the WFCL must make sure that its recommendations would have the support of the most powerful actors. The WFCL would have to rely on the market to provide the resources for research, the incentives for increased production, and the means for distributing the products.

By 1980 the WFCL was on record as supporting financial aid to famine or food-emergency areas. It had called for an elimination of protectionist trade practices. It had urged countries to reduce military budgets and devote the savings to the development of agriculture. It had reiterated the call for nationally held food reserves. And it had called for target commitments in money and in commodities to improve food security.[38] The WFCL's ability to achieve these goals has always depended entirely on the countries' willingness to commit themselves to national policies that would achieve them. The WFCL is limited in the same way FAO was. Members of FAO and the WFCL are countries. So the organizations do no more than the member countries are willing to do. Thus, it was logical for the WFCL to

---

World Food Conference's goals were: the World Food Council (WFCL), the Consultative Group on Food Production and Investment (CGFPI), and the International Fund for Agricultural Development (IFAD). The World Food Council would be made up of 36 countries and would meet at the ministerial level to coordinate the food policies of all the UN agencies. The International Fund for Agricultural Development's job was to marshall additional financial resources for food production. And the Consultative Group on Food Production and Investment was charged with bringing bilateral and multilateral donors together with representatives of developing countries in such a way as to coordinate resources for increased food production. By 1977 the UNDP, FAO, and World Bank had decided that the CGFPI's terms of reference either duplicated other agencies' responsibilities or were too broad to be attainable. The CGFPI was disbanded.

urge that the International Wheat Council (IWC) took on the task of coordinating the national grain reserve proposal called for by the World Food Conference because the IWC included all the major wheat producers. This multilateral attempt at managing food resources failed in spite of U.S. support, but the story of the failure provides insights into some changes in the climate of North-South negotiations, changes relevant to the prospects for managing food resources multilaterally.

The mostly Western members of the IWC[39] negotiated among themselves and arrived at a scheme for coordinating national reserve policies. All participating countries would agree to buy wheat when the price was low and sell it when the price was high. The negotiating was not without difficulties. Prices at which wheat would be bought and sold had to be established as did the methods for purchasing and releasing them. The interests of the IWC countries were not easily harmonized, but the big traders were able to arrive at acceptable technical means for achieving their common ends--price stability and food security. But the agreement they presented to the developing countries for approval was based on market principles. The developing countries rejected it.

> One interpretation of the closing drama is that the G-77 [Group of 77] wrecked a Western traders commercial deal which offered them little, thus demonstrating their power and depth of commitment to redistributive principles. Another more cynical version holds the LDCs were deluded into taking the blame for a failure to reach an objective which only they really wanted.[40]

As yet, no one has lost very much because of the IWC failure to establish a national food reserve system. "Following the collapse of the talks, the food aid convention guarantees were secured anyway, the IMF set up a special facility for compensatory food financing, and reserve stocks were built, including a U.S. four million ton food aid [wheat] reserve."[41]

At first it is difficult to understand why the developing countries would reject a plan that seemed to be in their interests. But their action is understandable if it is seen as a rejection of the IWC's market approach to a national reserve system. The Group of 77 was not rejecting the World Food Conference's goal; it was rejecting the whole notion of making simple technical adjustments in what would ultimately remain a market approach to food security and price stability. If the Group of 77 continues to reject proposals based on market principles, then the likelihood that any multilateral food resource management scheme will be widely accepted is virtually nil.

Given this impasse, it makes sense to simplify and disaggregate the global food problem somewhat. Perhaps some new basis for cooperation will appear if we consider the different requirements of the United States' different trading partners.

## PROBLEMS AND PROSPECTS FOR IMPLEMENTING THE MANAGEMENT OF FOOD RESOURCES MULTILATERALLY

There are four major food markets that the United States must understand in order to make sound policies for world food security and food price stability. Each of these markets has special requirements of its own that must be taken into account if cooperation on food issues is to survive the booms and busts in the market for agricultural products. The markets are: the European Community, Japan, the Soviet Union, and the developing world.

### The European Community

The European Community (EC) is an interesting case for U.S. policy makers. On the one hand, the Community supports, in principle, liberal trade policies. On the other hand, the EC spends billions to protect its agricultural sector from foreign competition. The problems of European agriculture are remarkably similar to those in U.S. agriculture. Europeans, particularly French farmers, produce too much, often in spite of somewhat inefficient methods. Prices for their commodities sink below what is needed to keep them in business. Without government support, without support from the European Community, the farmers would cease production altogether. So the Community supports agriculture because, among other things,[42] the European Community would be highly vulnerable to the instabilities and insecurities of the world market if it did not subsidize its agricultural sector. In times of scarcity it would be vulnerable to the kinds of political manipulations the Community experienced during the oil embargo.

In March 1984 the issue of farm subsidies became a crisis for both the farmers and the Community. Since the middle 1970s the farmers have been first encouraged by high prices and then required by low prices to produce more and more. During the scarcity of the 1970s, investments in agriculture looked more than reasonable; they were necessary if the countries of the Community were going to avoid the food price inflation experienced elsewhere. The Common Agricultural Policy (CAP), so controversial for so many years, protected the Community from the worst effects of the scarcity. But when the food glut returned and prices plummeted, the Community was forced by its earlier commitment to agriculture to pay more and more extraordinary subsidies to the farmers. The Community has been deadlocked on the issue of future payments because the British say they will not continue to subsidize inefficient European agriculture. The Community cannot just abandon its commitment to the farmers without losing its insurance against the insecurities and instabilities of world market conditions. To make matters worse, the individual farmers cannot afford to cut production with market prices so low. Each must squeeze every bit of production out of every input in order to survive the current crisis.

The reaction of the United States to the crisis in the Community has been to complain that the Community is illegitimately hurting U.S. farm interests by subsidizing and then exporting such products as wheat, sugar, and meat. As of March 1982 the United States was planning to file a formal complaint against the EC with GATT on the issue of flour subsidies. The United States was also considering complaints about citrus products from the Mediterranean that Europe imports under the Generalized Preference Scheme. The scheme allows some of the world's developing countries duty-free access to the EC market. The EC's typical response to U.S. complaints in GATT talks is to request something in return for allowing the United States to exploit its very considerable comparative advantage in agricultural products.[43]

The EC representatives point out that in spite of the decline in Europe's imports, the United States has managed to increase its exports from $6 billion worth in 1970 to $40 billion worth in 1982. The Europeans also point out that the United States did not ensure access to U.S. products during the shortages in the 1970s and that the Europeans therefore can hardly afford to rely on U.S. supplies in times of shortage. Nevertheless, U.S. farm interests pressure U.S. trade negotiators to break down the European barriers to U.S. products. U.S. Trade Representative, Ambassador William E. Brock III, testified before Congress that the United States was making a Cabinet-level, concerted effort to get the EC to lower its agricultural trade barriers and "to do it in a noncompensational fashion."[44]

The United States complains to the EC that U.S. interests are prejudiced by CAP policies, and this is undoubtedly true. But it is also true that the EC's policies are not to blame for all that is amiss in U.S. agriculture. Any approach to the EC that is likely to be effective would have to begin with the simple fact that the EC has a farm problem much like the U.S. farm problem and that the problem cannot be solved through market mechanisms alone for quite legitimate domestic and international reasons.

The EC's move toward self-sufficiency is not very efficient, but it is necessary for indigenous political reasons and for external, strategic reasons. When looked at in this way, EC policy does not look quite so unreasonable. It is in the United States' strategic interest that Europe be immune to food blackmail and that Europe be able to protect itself against economically damaging booms and busts in commodity prices. The United States has been able to substantially increase its exports in spite of CAP and, in general, the EC supports the United States' sales-oriented, free market approach to food security and stability. In the IWC negotiations, the United States and the EC found they had reconcilable differences with respect to each other and that their common adversary was the Group of 77. In fact, it looks as though it is the developing world that suffers most from CAP.

## Two Countries That Make Growing Demands on the International Grain Market: The Soviet Union and Japan

Both the Soviet Union and Japan have made increasing demands on the international grain market, both for roughly the same reason: They are increasing their consumption of meat. A country determined to produce more meat will need sixteen times as much land to produce chicken feed than it would to produce an equivalent amount of soy protein. Every acre devoted to animal feed produces one-third of the protein of an acre devoted to direct consumption of vegetable protein. Any multilateral attempt to improve food security and stability must factor in the effects of the trend toward greater meat consumption by these two important grain customers.

The USDA estimates that the Soviet Union alone "was responsible for 80 percent of all deviation from trend in world wheat imports." Paarlberg has called the Soviet Union "the world's largest and most irregular producer of wheat" and "the world's most disruptive grain trading nation."[45] After the 1972-73 fiasco the Soviets slowed the increase of new livestock production and in the bad weather years of 1975 and 1976 decreased the size of the livestock population. By the time of President Carter's grain embargo in January of 1980, the Soviets were able to adjust to the loss of U.S. supplies by buying from Argentina, Canada, Australia, and France. In the first year after the embargo was lifted, the Soviets bought about 14 million tons of wheat and corn and almost three-quarters of a million tons of soybeans from U.S. producers. The Reagan Administration has gone on record as opposing embargoes and even seemed hesitant to make another bilateral agreement with the Soviets establishing minimum and maximum purchases. Secretary Block testified before Congress that in the absence of any new agreement the Soviets would be able to purchase whatever they liked from the United States.[46] Neither the embargo nor its removal appears to have had any beneficial impact on the prospects for Soviet participation in multilateral efforts to improve food security and food price stability.

The United States has also made efforts to reassure the Japanese that there will be no new embargoes. The Japanese purchase 97 percent of their soybeans in the world market. Ninety-two percent of these soybeans come from the United States which produced 70 percent of the total world crop in 1973 when the United States stopped foreign sales of soybeans. Japan had allowed food self-sufficiency to decline when prices in the world food market were very low. In spite of expensive subsidies paid to rice producers and the fishing industry, Japan continues to depend on the world market for almost a third of its domestic food needs.

In spite of an inflationary and very expensive program to reduce Japan's reliance on other countries for its food supply, Japan is expected to make even greater demands on world supplies in the future. Japan's response to world market conditions has not been to lower meat consumption but to develop new sources of supply and to negotiate long-term bilateral deals with those sources.

## The Developing Countries

In his discussion of EC policy during the scarcity of the 1970s, Robert Paarlberg says:

> Having earlier dumped Europe's surplus onto a saturated world market, the CAP was now holding European produce away from a tight world market, in hopes of containing runaway food price inflation at home. All the while, the EEC continued to import substantial quantities of food. These trade practices raise questions about the overall contribution which European food policy makes to the security and stability of the larger world food system. . . .
>
> In certain respects the EEC takes a "free ride" on the world food system. In times of oversupply, the EEC made things worse by restricting imports and subsidizing exports; in times of shortage, the CAP has reduced EEC exports even while the EEC has continued to bid against less wealthy food deficit nations for a high volume of imports.[47]

The main point of Paarlberg's analysis is that the industrial food-importing nations have developed policies that protect them from the worst effects of food scarcity and glut while making those effects even worse for the developing nations. The United States and its allies in Europe and Japan advocate a loose multilateral solution based on the workings of the international market for grain. They argue that the most efficient strategy for developing countries is to accept the comparative advantage of the big producers and open their domestic markets to inexpensive foreign products. They argue that it is wasteful for these countries to duplicate agricultural production that is already excessive elsewhere. This argument is certainly compelling as far as it goes. Glut in the North matched with food deficits in the developing world ought to result in straightforward gain for all concerned. As we have seen, this formula has been around at least since the founding of FAO. Unfortunately, the problem is not so easily solved.

The developing countries need food price stability and food security even more than the developed countries because the tolerable margins for error are very much narrower and the consequences of miscalculation very much more grave. In the early 1970s when food prices soared in the United States, the EC, Japan, and the Soviet Union looked after their own interests. P.L. 480 allocations shrank, the EC held produce away from the market, and Japan and the Soviet Union simply bought what they needed. The Third World scrambled to adjust to these policies, but without resources or power, they had to settle for whatever assistance the developed world would spare. It was during this period that Earl Butz submitted a USDA budget without any allocation for P.L. 480 on the grounds that the problem for which P.L. 480 had been created, disposal of the U.S. surplus, had been solved. Remember also that Europe held back its agricultural products from the world market, stabilizing prices inside the EC but exacerbating the shortage in the world market. Remember also that the Soviet Union had precipitated the crisis by secretly purchasing huge grain supplies in the United States so that it would not have to reduce its livestock programs after the bad crop years. Third World countries learned a painful lesson in the 1970s. It was that when it came to scarcity of food, each country could expect little more than it could get fending for itself. The developing countries have begun to think more seriously about strategies of self-reliance.

The argument for self-reliance is that the efficiency attributed to comparative advantage is theoretical efficiency from the perspective of a global allocation of resources. Since the developing countries do not receive much advantage from the rationalization of resource allocation at the global level, it makes more sense, from their point of view, to withdraw as much as possible from the global exchange system.

Advocates of self-reliance make four interrelated arguments. First, dependence on the world market for food makes countries vulnerable to political pressures that can undermine their sovereign independence. U.S. officials speak openly about the "leverage" that food represents in U.S. dealings with developing countries. Agriculture Secretary Block cited at least two cases of U.S. leverage with respect to Egypt and Bangladesh, two countries with desperate food shortages.[48] While using this kind of leverage may seem reasonable enough from the point of view of U.S. policy makers, it is hardly ideal from the point of view of Third World leaders. Talk of using food as leverage to get the policy changes the United States wants in other countries is bound to encourage efforts for greater self-reliance.

Second, self-reliance would shelter developing countries somewhat from erratic market conditions and from capricious or malevolent market activity by other countries. Precious foreign exchange would not then have to be diverted from ongoing development projects in other sectors of the economy. India had to divert $600 million from other development projects to meet its food needs in 1983.

Indonesia spent $700 million. Even though some measure of efficiency would be sacrificed, self-reliant countries would not suffer all the inefficiencies that attend the reallocation of development resources when food prices change radically and without warning.[49]

Third, in developing countries an overwhelming proportion of the population lives in rural areas. Large, inexpensive imports of agricultural products undermine domestic producers. A policy of self-reliance encourages rural economic activity by sheltering it from the worst effects of foreign competition.

Fourth, developing countries, and the poorest people inside them, are rarely able to acquire enough money to create real, economically effective demand. Farmers produce when there is demand in the form of money. The market does not respond to need. Self-reliance may help fill the gap between the needs of developing countries and what the market will produce for its poorest customers.

In summing up the actions of the richest customers Paarlberg says, "each of these wealthy food customers prefers to tend to its own internal needs, and to force most adjustment burdens onto others."[50] In light of this well-known fact, it seems unlikely that the United States will be able to persuade very many countries that a free market in agricultural products is in their interest, in spite of the efficiency of comparative advantage.

## The Prospects for International Management

All of the developing countries combined do not import as much food as the European Community, Japan, and the Soviet Union. In these relatively wealthy countries meat consumption is rising and so are imports. All of them subsidize their domestic agricultural sectors to protect them from foreign competition. All of these countries receive a substantial portion of their imports from the United States. In fact, the United States has more than tripled its exports since 1960. The United States exported 130 million tons of grain in 1980, fully half of all the grain in international trade. Yet in spite of the expectation that increased exports of excess productivity would solve the farm income problem, the U.S. farm aid bill has grown in just three years to an astonishing $23 billion. The United States continues to produce much more than it can sell at a price that will keep U.S. farms in business. At the same time recent bad crop years in the developing world have put millions more people in jeopardy of starvation. On the face of it, one might think that there would be a very strong perceived need to collaborate. Yet the prospect of devising some institutional means for ensuring food price stability and food security seems more remote than when FAO was founded.

## THE FUTURE OF MULTILATERAL ATTEMPTS TO MANAGE FOOD RESOURCES

Scholars are sometimes overcome by the impulse to show how simply real world problems are solved when scholarly minds are brought to bear on them. The world food problem is too forbiddingly complex to tempt even the truly foolish. There are so many foreign and domestic interests to be reconciled and even as the policy makers are trying to reconcile them, each party's interest is changing because of changing economic and natural conditions over which no one has control. Under these circumstances, the best sort of contribution a scholar can make is to render a sound analysis of the situation confronting the policy maker. Scholars are not in a position to solve the problem, but they can show clearly what the nature of the problem is and say clearly what the policy makers should aim for. Perhaps it goes without saying that the policy makers' job is to do what is in the best interests of the country as a whole. They must decide whether any specific concession is worth making. But the scholar can make explicit what is often implicit or taken for granted: the standard the policy should meet. At present the two most important policy goals are better price stability and improved food security. These two are logically prior to the other goals because none of the other goals, freer markets, economic growth, more rational trade, sound aid programs, reduction of the farm aid bill, etc., can be achieved in the absence of greater food security and price stability.

If it is in fact true that security and stability are the appropriate goals, the next question is, what are the best means of achieving them? One thing that is quite clear from the history of the problem is that the United States cannot unilaterally ensure adequate levels of price stability and food security. Whether the United States embargoes grain sales to the Soviet Union or greatly expands its exports, the excess capacity of the United States cannot be disposed of in an orderly way without the cooperation of other nations.

It is true that the United States is not hurt as much as some other countries by the volatility of agriculture, but it is also true that no other country stands to gain as much as the United States from improvements in food security and price stability. The United States gains if it stabilizes the market because that is a precondition for stabilizing food prices, one of the important causes of domestic inflation. Market instability has also been the primary cause of the chaos in the agriculture market and the consequent huge farm aid appropriations of recent years. Market stability makes it easier for other countries to stabilize their own economies and as the premier status quo power, the United States benefits from the resulting political stability elsewhere. Food security is vital to every country's strategic security. It is simply shortsighted to think that food shortages, whether long- or short-term, acute or chronic, are consistent with political stability.

If stability and security are so vital to U.S. interests and if the United States cannot achieve these two goals through unilateral action, how can bilateral and multilateral cooperation of the kind that will serve U.S. interests be achieved? This is, of course, a very difficult question. But there are some clear, commonsense conclusions about what is likely to bring forth the necessary cooperation.

The first conclusion, and probably the one the free marketeers least want to hear, is that there is not going to be a free market in food at the international level. The United States may decide that it is worth the cost to negotiate very aggressively in GATT to maintain as much of the liberal trading system as it can. It is quite clear how much the United States benefits from the degree of free trade that now exists. But if other countries are asked to make "noncompensatory" concessions that might undermine their food security, erode their balance of payments position, and cause inflation and unemployment, the United States should expect them to refuse. Conditions endemic to agriculture make it extremely difficult for a market to function properly even within any particular country. Making it work on an international scale is not possible.

The second conclusion policy makers might draw from this analysis is that deliberate avoidance of political issues is unlikely to result in any real improvement on the issues of stability and security. The functionalists were wrong when they said that the interests were all the same when the problem was understood technically and scientifically. The mere adoption of nutritional standards has not had a beneficial effect on agriculture and trade because ignorance of nutritional need was not the obstacle to cooperation. The obstacle to cooperation was and is the fact that different countries have different interests.

The big producers and the EC want a free market allocation of resources. They know they will benefit more from this policy than the developing nations, but they say that eventually the Third World will benefit because international trade will be stimulated and economic growth will occur. The greatest good for the greatest number will be achieved when the most efficient producers are allowed to pursue their advantage by driving less efficient producers out of business and reallocating their inefficiently used resources to more efficient producers.

The developing countries want to debate the reasonableness of the free market allocation of resources. They say that free markets consign the developing countries to permanent inferiority. The already wealthy are in a better position to do well in the market place because money is power in the market. Countries with the most food and the most money will be able to make better deals. They have the option of buying food only when prices are low and selling only when prices are high.

The developing countries assert that the market allocation is not moral because the market allocates disproportionate resources to the rich when the poor are starving and lack even the minimal resources to create effective economic demand.

The most important obstacle to the Group of 77 accepting the IWC proposal was its embodiment of free market principles. The developing countries, for all the reasons mentioned above, were wary of the market. As the IWC negotiations so clearly showed, the developing countries are willing to hold out for a redistributive, multilateral food reserve system. If they continue to hold out, the United States will have to enter into a debate on the adequacy of market allocation. If the United States wants to persuade the developing world that the market will work, it must show exactly how the market can work in the developing countries' interests.

It is unlikely that such a debate will be conclusive, but what the United States can usefully do in the meantime is to set specific attainable policy goals that any reserve system should achieve. The United States may have a very strong interest in a market system, but it must also find ways to reduce the farm aid budget, insulate the U.S. economy from the worst effects of erratic markets, and remove the politically destabilizing effects of chronic hunger. These three goals may be attainable in ways that do not compromise the U.S. commitment to the market. Scholars are not in a position to say precisely what the resulting reserve system would look like. Too much depends on what happens in specific negotiations. But scholars can caution policy makers against insisting so much on a market solution that they miss chances to reduce the farm aid budget, stabilize prices, and develop innovative proposals to reduce chronic hunger.

## NOTES

1. Robert D. Hormats, Assistant Secretary of State for Economic and Business Affairs, in testimony before the 97th Congress, March 9, 1982.

2. Otto Doering, Andrew Schmitz, and John Miranowski, The Full Costs of Farm Exports (U.C. Berkeley: Giannini Foundation of Agricultural Economics, June 1982), p. 15.

3. I. M. Destler, "United States Food Policy 1972-1976: Reconciling Domestic and International Objectives," in The Global Political Economy of Food ed. Raymond F. Hopkins and Donald J. Puchala (Madison, Wisconsin: The University of Wisconsin Press, 1978). I wish to acknowledge my indebtedness to Destler's analysis. His article greatly influenced the argument made here that there are lots of legitimate interests to be served in the matter of food and agriculture policy making.

4. Ibid., pp. 46-47.

5. Raymond F. Hopkins and Donald J. Puchala, "Perspectives on the International Relations of Food," in The Global Political Economy, p. 7.

6. Destler, ibid., p. 58.

7. For evidence of instability due to food prices, see Nicholson and Esseks, in ibid., pp. 114-120.

8. Destler, ibid., p. 58.

9. Ibid., p. 59.

10. Hormats, testimony, p. 10.

11. Gove Hambidge, The Story of FAO (New York: Van Nostrand, 1955), p. 48.

12. Ibid., p. 49.

13. Ibid., p. 50.

14. Ibid., p. 50, emphasis added.

15. John D. Black, ed., Annals of the American Academy of Political and Social Science, vol. 225 (January 1943). See also Raymond J. Sontag, A Broken World: 1919-1939 (New York: Harper Row, 1971).

16. For a poignant description of the demise of the League and the establishment of the UN, see Leland M. Goodrich, "From the League to the United Nations" in the first issue of the journal International Organization. This article is reprinted in International Organization: Politics and Process, ed. Leland M. Goodrich and David A. Kay (Madison, Wisconsin: University of Wisconsin Press, 1973), pp. 3-21.

17. Inis L. Claude, Swords into Plowshares: The Problems and Progress of International Organization (New York: Random House, 1971), Chapter 4 and Part 3.

18. Ernst B. Haas, Beyond the Nation-State: Functionalism and the International Organization (Stanford, California: Stanford University Press, 1964), pp. 120-25 and 537. Haas includes Gunnar Myrdal at the UN Economic Commission for Europe, Julian Huxley at UNESCO, Brock Chisholm at the World Health Organization, Eugene Black at the World Bank, Albert Thomas at the International Labor Organization, and key figures at the UN Commission for Latin America, UNRRA, the Special Fund, the Coal and Steel Community, and Boyd-Orr at FAO.

19. Robert W. Gregg, "UN Economic, Social, and Technical Activities" in The United Nations: Past, Present,

and Future, ed. James Barros (New York: Free Press, 1972), pp. 222-23.

20. Hambidge, The Story of FAO, p. 55.

21. Norris Dodd, second Director-General of FAO. In ibid., p. 87.

22. Hambidge, quoting from The Work of FAO, p. 55.

23. Black, Annals of the American Academy, p. 123.

24. Ibid., p. 124.

25. Stephen S. Goodspeed, The Nature and Function of International Organizations, 2nd ed. (New York: Oxford University Press, 1967), p. 443.

26. Haas, Beyond the Nation-State, p. 537.

27. Ibid., p. 122.

28. Black, Annals of the American Academy, p. 127. Emphasis added.

29. Hambidge, The Story of FAO, p. 52.

30. Ibid.

31. Hambidge, The Story of FAO, p. 52. In Beyond the Nation-State, Haas says "exporting nations were not ready to consider an international commodity policy," p. 122. Boyd-Orr describes the failure of his plan to buy surplus commodities and sell them to people who needed them, thus stabilizing prices and improving nutrition simultaneously. John Boyd-Orr, Man's Dilemma (London: George Allen and Unwin Ltd., 1964).

32. Hambidge, p. 49. See also Goodspeed, for Article 1 of the FAO constitution, p. 433-34.

33. Boyd-Orr, p. 57.

34. Haas, p. 122.

35. Hambidge, p. 88.

36. Ibid. Quoting a UN resolution. Emphasis added.

37. Quoted in Thomas G. Weiss and Robert S. Jordan, The World Food Conference and Global Problem Solving (New York: Praeger, 1976), p. 11.

38. For a complete review of the WFCL's activities see Comptroller General of the United States, World Hunger and Malnutrition Continue: Slow Progress in Carrying Out

World Food Conference Objectives (Washington, D.C.: U.S. General Accounting Office, January 11, 1980).

39. The key countries in these negotiations were the United States, the European Economic Community, Canada, Australia, Argentina, and Japan. This account of the IWC negotiations is taken from Raymond F. Hopkin's unpublished paper, "The Wheat Negotiations: Loss or Gain in North-South Relations?"

40. Ibid., p. 17.

41. Ibid., p. 38.

42. Among the most important "other things" is the fact that the French made support for their agricultural sector a condition of their going in the community.

43. Mitchell B. Wallerstein, "Foreign-Domestic Intersections in U.S. Food Policy," Food Policy 5 (May 1980):88.

44. William E. Brock III, testimony before the second session of the 97th Congress, March 9, 1982.

45. Robert L. Paarlberg, "Shifting and Sharing Adjustment Burdens: The Role of the Industrial Food Importing Nations," in The Global Political Economy, pp. 96-97.

46. U.S. Congressional Committee on Agriculture, p. 93.

47. Paarlberg, "Adjustment Burdens," p. 86.

48. Hearings before the House Committee on Agriculture on "General Agricultural Export and Trade Situation," 97th Cong., 2d sess. (Washington, D.C.: Government Printing Office, 1982), pp. 6-7. See also Hearings before the House Committee on Foreign Affairs, "Foreign Assistance Legislation for Fiscal 1983," 97th Cong., 2d sess., pp. 6 and 16-17.

49. See Bruce F. Johnston and John W. Mellor, "The World Food Equation: Interrelations Among Development, Employment, and Food Consumption," Journal of Economic Literature 22 (June 1984).

50. Paarlberg, "Adjustment Burdens," p. 101.

# PART III

## THE UNITED STATES AND THE MULTILATERAL MANAGEMENT OF MARINE MINERAL RESOURCES

# 5

# THE INTERNATIONAL SEABED AUTHORITY: PROGRESSIVE DEVELOPMENT OR CREATIVE CAMOUFLAGE?

*Roger A. Coate*

## INTRODUCTION

Arvid Pardo's impassioned plea on August 17, 1967 to ". . . declare the seabed and the ocean floor a common heritage of mankind" (UN Document A/6695) set in motion, both within and outside the United Nations, a complex and fascinating chain of events. In the decade and a half that followed, governmental and nongovernmental participants alike struggled with the task of operationalizing such a far-reaching principle. From the beginning the idea of creating an international institution to fulfill such a role served as a focus for these "seabed regime" activities. By the end of the Third United Nations Conference on the Law of the Sea (UNCLOS III) in December 1982, a near-consensus had been reached on the creation of such an International Seabed Authority (ISA) within the context of a much larger Law of the Sea "package" (Convention on the Law of the Sea, UN Document A/CONF.62/122).

The nature of this ISA and its implications for the operationalization of the common heritage principle are the main foci of this chapter. More specifically, the discussion below focuses on the extent to which such an authority represents progressive development within the international legal order toward the ends of benefiting all mankind or conversely merely represents another variant within the status quo legal order. To provide a basis for such an assessment, the discussion below focuses on three interrelated tasks. First, the evolution of the ISA as an international institution is examined. Specific attention focuses on the deliberations which occurred within the United Nations Seabed Committee and Committee I of UNCLOS III. Second, the nature of the larger global social order within which an ISA would have to operate is described in a manner which illuminates various structural parameters inherent in the global political economy. Third, the implications of the seabed exploitation provisions of the Convention on the Law of the Sea are examined in light of such structural concerns. The chapter then concludes with an overall assessment of the role of the seabed exploitation regime in providing for the common heritage of mankind.

## THE EVOLVING SEABED REGIME

In his call for declaring the international seabed to be considered a common heritage of mankind, Pardo had envisioned the establishment of an international body to control, regulate, and supervise the exploration and exploitation of the seabed and ocean floor beyond the limits of national jurisdiction. Pardo's call almost immediately sparked a debate over the nature of such international machinery and control. Thus, from the beginnings of seabed deliberations in the United Nations, this concern became a major issue.

During this formative period delegates' views tended to vary within relatively narrow limits. On one extreme, the preference was articulated that no formally institutionalized seabed authority should be established. Such a view was held by the Soviet delegate and several other East European delegations. At the other extreme, numerous delegates from nonindustrial areas tended to support what would later be considered a rather moderate "mixed" system of exploitation. Under such a "mixed" system private and state enterprises could exploit the Area under license by an international seabed authority, which would also undertake mining operations.

An "intermediate" position was taken by most delegations from industrialized market economies. These delegates tended to argue for a very limited seabed authority. The general preference among this group was in support of a *quota* licensing system. Under such a system the international seabed authority would serve basically to facilitate private and state mining operations through license allocation activities (allocation being based on a national quota system). The British, French, and Japanese delegations were the main proponents of such an approach.

The position of the U.S. delegation differed somewhat from such a general intermediate position. The 1970 U.S. proposal for a Draft Convention (A/AC/138/25 [1970]), while also favoring a licensing approach, called for open and nondiscriminatory access to the Area. A national quota system could discriminate against the more numerous U.S.-based potential seabed miners by limiting the number of licenses available to such firms.

Like the other intermediate positions, however, the international seabed authority envisaged in the U.S. delegation's proposal could be described as a minimal control system. An important aspect to note, given our present interests, is that in the pre-1971 period such a limited authority approach represented the middle point on a continuum of proposals regarding the question, "Who may exploit the area?" During this period delegations seemed to be orienting their proposals and arguments more toward what was perceived as being achievable (given internal capabilities and constraints) than what was ideally desired or politically and strategically wise.

The three draft proposals had emerged during the Seabed Committee deliberations in 1970. These proposals were introduced by the delegations from the United States (A/AC.138/25), the United Kingdom (A/AC.138/26), and France (A/AC.138/27). Each had represented an intermediate position within the context of the general preferences expressed by most delegates during the first three years of the Committee's deliberations.

However, as preparations began in 1971 in Subcommittee I for the formulation of a set of draft treaty articles to be presented for consideration at UNCLOS III, a much wider range of proposals emerged.[1] These proposals represented a much broader range of alternative positions than had the earlier set of proposals and committee discussions. A very general and limited set of complementary notions was characteristic of most of these proposals. In general, these ideas included the notions that: some kind of international machinery would be created; that the 1970 Declaration of Principles resolution (2749[XXV]) should serve as the basis upon which the draft articles were based; and that, while observing the special needs of governments from peripheral areas, the revenues from seabed exploitation should in some way benefit all states. Beyond such general and limited concerns, however, substantial disagreement emerged.

For example, the Tanzanian proposal (A/AC.138/33, Article 13) represented the first formal proposal calling for a mixed system of exploitation and production controls. These notions ran quite counter to the draft articles supported by the delegations from industrialized areas. Controversy emerged over the nature and scope of the ISA ("Who may exploit the area?") and the nature and distribution of functions and powers among organs of the ISA (Council, Assembly and Enterprise). Three major groups served to focus the deliberations in Subcommittee I: the Group of Five; the Group of 77; and Landlocked and Geographically Disadvantaged States (the Disadvantaged Group). On this issue, though, little actual negotiation occurred across groups.

At the end of the Subcommittee's work a set of draft articles was produced, which reflected various conflicting positions as well as points of agreement. This (Draft) Convention on the Seabed and the Ocean Floor Beyond the Limits of National Jurisdiction was to serve as a basis for negotiations within the First Committee at UNCLOS III (A/CONF.62/C.1/L.3 [1974]). The cleavages that were inherent in the alternative article provisions specified within the document reemerge in a much more intensified fashion at UNCLOS III.

The story of seabed negotiations within UNCLOS III is very much one of the evolving nature of an intergovernmental political movement within the so-called "South" (peripheral areas). While the proposal (the 13-Power Proposal) calling for a strong, centralized ISA and Enterprise drew only limited support within Subcommittee I of the Seabed Committee, substantial support developed rapidly at the conference. By the end of 1975 such an idea had captured the support of an overwhelming majority of conference participants.

Underlying the challenge presented by the 13-Power Proposal was a basic conflict over the definition of the nature of the issue(s) in contention. To many of the participants from peripheral areas a legal principle was at stake--the common heritage of mankind. This new principle was to serve as a basis for redefining international public law and relative positions of national governmental participants within the global social order. Entailed within such a redefinition would be a basic redistribution of tangible benefits from the exploitation of global "commons."

To governmental bodies from the industrialized areas such statements represented only value premises, not legal principles.[2] The political goods at stake were perceived as being primarily tangible to these participants. For them seabed exploitation "rights" could be utilized. Commercial and/or state-owned interests they represented were in a position to seriously consider mining minerals from the seabed. A system for regulating such mining operations might be acceptable, even beneficial. However, the notion of having such operations fall under the exclusive control of an ISA was totally unacceptable. Also, while these participants might be willing to distribute proceeds from licensing and other revenues of the ISA to less affluent areas, the idea of redistributing seabed mining technology to the Enterprise or to governments from peripheral areas was strongly opposed.

At Caracas in 1974 a new important element entered the debate: the economic implications of seabed exploitation. Governments from peripheral land-based mining areas were especially anxious over this question.[3] It was perceived that active seabed mining might seriously jeopardize export earnings from land-based mining operations in respective mineral industries. The important aspect of this question for later developments within UNCLOS III, though, was the bridging function that this concern created between seabed questions and larger commodity questions related to the call for the establishment of a New International Economic Order (NIEO). Such larger concerns were in turn linked to more ideological questions related to global inequalities.

The lines of opposition within the First Committee were, at this point, becoming clearly drawn between the Group of Five and the Group of 77.[4] The text on "basic conditions" introduced by the Group of 77 did move somewhat in the direction of providing specificity, as demanded by the U.S. delegation. However, ". . . the more the emphasis was put on details in Committee I the further the groups moved apart from each other and the more ideology rather than economics was the major variable."[5]

The dominant strategy which emerged within UNCLOS III by various delegations within the Group of 77 was that of issue linkage. Two general types of linkages were attempted: (1) linking seabed issues with issues of global inequalities; and (2) linking concessions on other oceans issues with concessions on seabed issues.[6] The first type of linkages served to build cohesiveness within the group. Such a strategy, however, tended to push the two sides apart. The second linkage strategy worked in the other direction, attempting to win concessions with respect to the establishment of a strong and meaningful ISA. By linking the demand for a strong seabed "authority" (on which the major industrialized, maritime governmental delegations were split internally) to matters on which those delegations are deeply interested and more unified, these relatively weak delegations have been fairly successful in getting the larger governments to move closer to the Group of 77 position.[7]

The strategies of the delegations from the Group of Five, most particularly the U.S. delegation, were much more oriented toward the direct and indirect application of influence capabilities.[8] Given the concentration of ocean mining technology and financial capital in core areas and given the conference decision-making structure under the Rules of Procedure, these delegations were operating from a relative position of strength.

Such a statement, however, must be considered in light of institutional requirements and constraints within which these UNCLOS III delegates were operating. Such requirements and constraints played a major role in shaping the general nature and direction of seabed negotiations. For example, it was widely recognized that officials in the U.S. Department of Defense strongly supported the signing of an international treaty that would limit jurisdictional claims over the high seas. Given the demonstrated influence on this issue of that department within the U.S. Executive Branch since 1970, various delegations from the Group of 77 perhaps accurately perceived a chance to bargain for concessions in exchange for such an agreement. On the other hand, however, there were numerous participants within the U.S. government who wanted no treaty at all, if such a treaty contained constraints on seabed mining operations. Also, there were strong pressures for the passage of U.S. unilateral seabed mining legislation. This latter factor provided the U.S. delegation with a very important implicit influence capability.[9] This factor, along with the passage of the Fisheries Conservation and Management Zone Act (PL 95-264) and the growing threat of unilateral seabed legislation enabled the U.S. delegation to press for major changes in what U.S. officials perceived to be a very unacceptable Informal Single Negotiating Text (SNT), which had emerged at the end of 1975. The Revised Single Negotiating Text which appeared at the end of the first session of 1976 was much more palatable to the U.S. delegation than had been the SNT.

During the 1975 and 1976 seabed negotiations, the U.S. delegation also made major changes in its negotiating positions with respect to the nature and scope of the ISA. In 1975, following a Soviet delegation proposal for a "parallel system" of seabed exploitation, the U.S. delegation proposed what was termed a "banking system." Under this latter plan 50 percent of the Area would be open to exploitation by state or private mining enterprises, under "basic conditions" to be spelled out in the treaty. The remaining 50 percent would be reserved for exploitation by the Enterprise.

This proposal was followed on April 8, 1976, by a speech by Secretary of State Henry Kissinger in which he stated that the U.S. government would accept a parallel system in which "prime mining sites are reserved for exclusive exploitation by the Enterprise or by the developing countries directly."[10] He also stated that

> The United States is prepared to accept a temporary limitation, for a period fixed in the Treaty on production of the seabed minerals tied to the projected growth in the world nickel market. . . .[11]

Given previous negotiating positions of the U.S. delegation, these statements represented a substantial shift.

Indeed, this shift represented a major element in the development of the seabed regime and in the successful conclusion of UNCLOS III. By 1980 a remarkable "package" of compromises had been negotiated. The major areas of consensus within this package include a three-tiered voting structure for the Council, a seabed production policy, a technology transfer policy, an Enterprise financing arrangement, a two-stage approval process for mining applications, and a general policy on the distribution of benefits from seabed exploitation. These various compromises were in turn incorporated along with other revisions into a Draft Convention on the Law of the Sea (A/CONF.62/WP.10/Rev.3).

The participants within UNCLOS III had come remarkably close to overall agreement. Basically four relatively small issues remained unresolved at that point. These included questions related to: (1) participation (Can entities other than national governmental bodies become parties to the convention?); (2) the nature of the preparatory commission (matters relating to participation; decision-making procedures; financing and termination); (3) maritime boundaries of opposite or adjacent coasts; and (4) preparatory investment protection (PIP) for seabed mining operations undertaken before the convention comes into force.

Success of the conference seemed well within the realm of possibility. Delegations had agreed in August of 1980 to conclude a treaty in early 1981. However, in March of 1981 the new U.S. president decided to conduct a comprehensive review of the U.S. delegation's position within the conference and not to conclude negotiations on a Law of the Sea Treaty.

At the conclusion of the U.S. policy review, a set of proposed detailed amendments (the "Green Book") to the Draft Convention was submitted. Among the more unimportant changes demanded by the U.S. delegation were the demands that the United States be assured a seat on the 36-member Council of ISA and that decision making in that body be based on a weighted voting scheme, reflecting members' "political and economic interests" in seabed mining activities. Also, under the proposed changes U.S.-based mining enterprises would be guaranteed access to seabed minerals and would have no legal obligation to transfer technology to the Enterprise. The idea of a production ceiling to protect land-based producers was similarly rebuked, as was the possibility that the PLO could become associated with the convention. In addition, a structural problem was cited with regard to the treaty provision that allowed amendments to be made to the convention by two-thirds majority vote at the review conference which was scheduled to be held after 15 to 20 years. Thus, binding changes could be made in the text of the convention without the formal consent of the U.S. government.

By the time the final conference vote was taken at the request of the U.S. delegation on April 30, 1982, the Reagan Administration had determined that sufficient progress had not been made to resolve these difficulties. Along with Turkey, Israel, and Venezuela, the U.S. delegation voted against the text of the convention.[12] Most European seabed mining states, including Belgium, the Federal Republic of Germany, the Netherlands, and the United Kingdom abstained on the vote, as did the Eastern European delegations. The delegations of Austria, Australia, Canada, France, Greece, Ireland, Japan, New Zealand, and the Scandinavian states voted in favor, as did most delegations from developing countries. Thus, with a final vote of 130 in favor, 4 against, and 17 abstentions, the conference adopted the United Nations Convention on the Law of the Sea.

On December 10, 1982 the UNCLOS III was opened for signature. A total of 119 delegations, including 117 states, signed the treaty on the first day. The agreement will come into force after ratification by 60 states.

The signing of the UNCLOS III Treaty did not mean an end to the multilateral negotiations over seabed issues. Numerous technical and procedural questions related to bringing the ISA and the Enterprise into existence remained. A preparatory commission had been established and assigned the task of resolving such concerns. This body began work in March 1983 to carry out such a mandate. Indeed, the foundation upon which to establish the ISA seems to be well in place.

But how does one assess the potential utility of such institutional arrangements for operationalizing the common heritage principle? One obvious step that has to be taken is to examine the various constraints within which an ISA must operate. In so doing the nature of the principle, itself, will be substantially clarified.

Any seabed regime institutional mechanisms must operate within the confines of the much larger global social order. Given the nature of seabed exploitation activities, the global political economy is of particular relevance. The task, then, becomes one of understanding the nature of the structural interrelationships within such a system.

## THE GLOBAL POLITICAL ECONOMY

Any attempt at clarifying the nature of such global structural interrelationships must focus major attention on the capitalist world economy. Immanuel Wallerstein has traced the origins of such a "world-system" back to the sixteenth century. During this period, he argues, there emerged in Europe a "world" political economy distinct from previous forms of multi-societal organization.[13] According to Wallerstein's thesis, the dominant characteristic of this world-system was that it was based on a capitalistic mode of production. A central defining characteristic of such a political economy is the production of goods for sale for profit in a market. During this historical period this capitalistic world-economy was centered primarily around agricultural production, involving a multi-societal market structure. Such a political economy had its foundation in a system of ownership and control over production processes and a resulting unequal division of labor. A producer combines productive factors (resources) to create value in excess of factor input and other production costs. Accordingly, a key to growth within the system is the capacity to expand the scale of and to accumulate capital.

Financial capital plays a key role within a capitalist world economy. In addition to serving as a store of wealth, it also serves as a unit of account and medium of exchange. Financial capital, then, serves as a medium through which other resources can be acquired or transformed into one another. Accumulated capital can be reinvested back into production processes (in the form of new technology and capital goods), creating growth within the firm and specific areas of the political economy and, thus, promoting the further accumulation of capital. While being a necessary element delimiting growth, capital alone is not sufficient. A closely related element is technological change of production processes and marketing capabilities and the resulting capability to expand factor and product markets.

Technology serves dual purposes within such a process. It is a resource defining production technique and, in addition, serves as a basis for creating new resources. Thus, control over technology (i.e., the ability to create knowledge regarding new and/or more "efficient" combinations of factor components in production) is also an important consideration for understanding the evolution of a capitalist world economy. As the relative costs of various factor components change over time, the producer needs to be able to readjust production processes, increase revenues from the product market, or innovate some other means of maintaining the capability to accumulate profits. It is in this general context that concentration of control over capital, technology, and institutional mechanisms through which market structures can be affected becomes a critical element within a global political economy.

The notion of technology within the capitalist world economy (or any other social context) is not value free, however. Deeply ingrained in technology are assumptions about social organization and other values (e.g., the organization of knowledge, social structures, human relations with the geophysical environment and so forth). Johan Galtung argues that every technology carries with it its own economic, social, cultural, and cognitive "codes" for structuring human behavior.[14] Thus, technology delimits choices regarding the development and structure of social organization. In a capitalist world economy, the notion of "modern technology" tends to be employed synonymously with production techniques which are capital intensive, research intensive, energy intensive, organization intensive and labor extensive.[15] Thus, there emerges a distinct division of labor within such a system.

## Global Division of Labor

This division of labor is based on a complex system of unequal exchange relationships, both at micro- and macro-levels. For example, labor not only serves as a factor input in production, but laborers also often serve as consumers of finished goods. Thus, they not only provide their labor to the production process for what is often a nominal wage but, in addition, transfer wage capital back to owners of production processes and factors.[16]

Not only is the capitalist world-economy characterized by an unequal division of labor involving the appropriation of surplus value from certain participants to other participants, but also the appropriation of surplus value within the world-system from certain areas or zones to others. In Wallerstein's thesis this latter division of labor is the fundamental one.[17] He argues that in order to maintain a stable and enduring capitalist world-economy a three-fold division of labor is required. Thus, he distinguishes between core, semi-peripheral, and peripheral areas (zones) within the world-economy. The position occupied by a given area, according to Wallerstein, is to be explained by the structural role played by that area within the larger world-economy at a particular point in time.[18]

According to Wallerstein's thesis, "the semi-periphery is needed to make a capitalist world-economy run smoothly."[19] Two reasons are posited regarding such a "need."[20] First, the presence of semi-peripheral areas keeps the system from becoming highly polarized "with a small distinct high-status and high-income sector facing a relatively homogeneous low-status and low-income sector . . ." The second reason stems from the discontinuity between economic boundaries and states' jurisdictions, which Wallerstein argues constitutes the singular most important distinguishing feature of the capitalist world-economy.

> The multiplicity of states within the single economy has two advantages for sellers seeking profit. First, the absence of a single political authority makes it impossible for anyone to legislate the general will of the world-system and hence to curtail the capitalist mode of production. Second, the existence of state machineries makes it possible for the capitalist sellers to organize the frequently necessary artificial restraints on the operation of the market.[21]

However, such a "multi-state" system within the world-economy has an important disadvantage for sellers. Forces other than those controlled by sellers also can exert influences through governmental machinery. Thus, laborers' abilities to influence processes of authoritative allocation of values, and thus to increase wage levels within core societies, serve as constraints on production activities within such societies. Wallerstein goes on to argue

> The rising wages of the workers in the core countries, combined with the increasing economic disadvantage of the leading economic producers, given constant technological progress, and heaviest investment in rapidly outdated fixed capital by precisely the leading producers, leads to an inevitable decline in comparative costs of production. For individual capitalists, the ability to shift capital, from a declining leading sector to a rising sector, is the only way to survive the effects of cyclical shifts in the loci of the leading sectors. For this there must be sectors able to profit from the wage-productivity squeeze of the leading sector. Such sectors are what we are calling semi-peripheral countries.[22]

Thus, Wallerstein sees semi-peripheral areas as being necessary for preventing major crises within the system.[23]

Wallerstein's careful and thoughtful historiography of the evolution of the capitalist world-economy provides us with an interesting analytical tool. His thesis helps us to understand better the nature of concentration of control over capital, technology and various other resources that are critically important in most social exchange relationships. The three-fold structural division of labor can be viewed as representing differing "resource zones" within which participants are structurally situated. These zones can be conceived of as unequal resource environments which provide important requirements and constraints on global social interactions.

## Patterns of Global Resource Concentration

Control over financial capital, technology, and various other important resources within a global political economy tends to be heavily concentrated in core areas. The global activities of transnational corporations are indicative of such concentration. Such activities are highly concentrated both in terms of geographic distribution and industrial sector. Three-fourths of all direct foreign investment is located in developed market economies. Of the total estimated stock of foreign investment, over half is owned by the United States. When British, French, and German ownership of such investment stocks are added on, the proportion exceeds four-fifths.

> Moreover, foreign direct investment tends to be concentrated in a few firms within each home country. For the United States, about 250 to 300 firms account for over 70 percent. For the United Kingdom, over 80 percent of the total is controlled by 165 firms. For the Federal Republic of Germany 82 firms control over 70 percent, and the nine largest foreign investors alone control 37 percent of the total.[24]

Not only is the origin of such investment highly concentrated within core areas, but so also is the geographical destination. Three-fourths of all direct foreign investment is located in industrialized market economies. It is with respect to the remaining fourth, however, that a clear picture of the intensity of resource concentration within the global political economy emerges.

U.S. transnationals account for over half of the total stock of direct foreign investment in the so-called Third World. About 70 percent of such U.S. investment has been concentrated in Latin America. In turn, in Latin America direct foreign investment by U.S.-based transnationals accounts for two-thirds of the total stock of all such investments.

> In Africa, on the other hand, the United States accounts only for one-fifth of the total stock; the United Kingdom predominates with 30 percent, France following with 26 percent. Belgium, the Netherlands and Italy account for 7, 5 and 4 percent respectively. In the Middle East, the United States accounts for 57 percent, the United Kingdom for 27 percent, and the Netherlands and France for approximately 5.5 percent each. In Asia, the United Kingdom has the largest share (41 percent), the United States follows with 36

> percent, France with 7 percent and the Netherlands with 5 percent.[25]

Such regional concentration figures tell only a small part of the story, however.

Direct foreign investment in underdeveloped regions has tended to be concentrated in only a relatively few national economies. In the mid-1970s, 36 percent of all such investment was located in five national economies: Brazil, Mexico, Venezuela, Indonesia, and Nigeria. Nine others (India, Malaysia, Argentina, Singapore, Peru, Hong Kong, the Philippines, Trinidad and Tobago, and Iran) accounted for another 22 percent.[26] Thus, over half of all such transnational corporate activity was concentrated in only 14 national economies.

Concentration is further intensified by the fact that "this concentration is related to the sector in which foreign investment is predominant."[27] In general, in Africa, Latin America, and the Middle East such investment was heavily concentrated in extractive industries. In particular national economies (i.e., Argentina, Brazil, India, Mexico, and the Philippines and Singapore) direct foreign investment in manufacturing was intense. In both Indian and Malaysian national economies, such investment in agriculture was heavy.[28] Such investment generally "accounts for important percentages of the total stock of local investment, local production and sales."[29]

In addition, transnationals tend to cluster in industries which are dominated by a few large firms (e.g., extractive industries, chemicals, transportation, food products, and machinery). It is precisely in such industries that transnationals find the conditions that are so conducive to direct foreign investment: monopolistic advantage. The advantage to be exploited is generally sophisticated technology, linked with economies of scale, vertical integration, and global marketing capabilities. The U.S., British and, above all, the German transnationals, tend to concentrate in high technology industries, within which they control accesses to the technology. Such control, when linked with heavy advertising, global operating scale and product differentiation, makes it exceedingly difficult for local firms to compete successfully with the transnationals.

Joan Spero provides a concise overview of patterns of direct foreign investment in two Latin American national economies.

> In both Mexico and Brazil foreign investment plays a pivotal and growing role in the economy. In Mexico multinational corporations produced 23 percent of the total gross domestic product in manufacturing in 1970, up from 18 percent in 1962. In 1972 multinational corporations accounted for 28 percent of all Mexican sales, up from 20 percent in 1962. In Brazil in 1974 foreign multinationals controlled 29 percent of all assets in the manufacturing sector, and produced

> 44 percent of all local sales in the manufacturing industry in 1975.[30]

She goes on to discuss the substantial foreign control over the largest corporations in key industries:

> In Mexico 50 percent of the three hundred largest industrial firms and 61 percent of the hundred largest firms were foreign controlled in 1972. Foreign control is particularly predominant in the most technologically advanced and capital-intensive industries in Mexico: for example, in non-electrical machinery (95 percent), transportation (79 percent), and chemicals (68 percent). . . . Finally, foreign investment is found in highly concentrated industries. Multinational corporations accounted for 71 percent of all manufacturing sales in highly concentrated industries.[31]

The data for Brazil demonstrated similar patterns.

The reason that the Mexican and Brazilian cases are so important and enlightening is that these two national economies are among a very small group of peripheral societies that have been labelled by some observers as newly industrializing countries (NICs).[32] Such peripheral national economies have been singled out because of their relatively high rates of growth. Yet in both Mexico and Brazil the fastest growing and most profitable industries are dominated by transnational corporations. Such sectoral domination by foreign enterprises also manifests itself in the form of patenting practices.

Restricting access to technological know-how through licensing is one way of attempting to maintain monopolistic control. With respect to transnationals operating in underdeveloped areas, it has proved to be a very effective way. In peripheral areas an overwhelming number of the patents granted are held by nonnationals. It is estimated that ". . . if the number of patents is weighted by their economic or technological worth (i.e., volume of sales or value added) most developing countries are likely to find that the so weighted patents belonging to their own nationals amount to a fraction of 1 percent of the total patents granted by such countries."[33] The question of the role of patents in peripheral areas is linked to a much larger concern labeled by Lynn Mytelka as the "technological dependence syndrome."[34] By clustering in industries with "complex and volatile" technologies, transnationals' technological superiority and control permit the extraction of monopoly rents. In a study of metal working and chemical firms in Peru, Ecuador, and Colombia, Mytelka found that when foreign-owned or even state-owned corporations were involved, they would overwhelmingly acquire technology through foreign-owned licenses.[35] Such was the case even when relatively simple technology was involved. The study also revealed that such corporations tended not to develop local research and development capabilities. "Licensing itself becomes a factor inciting and maintaining technological dependence."[36]

In general, the nature of transnational corporate activity within peripheral areas has changed in the last half century. The traditional pattern of such activity entailed the investment of external capital in peripheral areas to produce goods to be exported back to core areas. Such "enclave" types of foreign investment have been described thus:

> In enclave economies, foreign invested capital originates in the exterior, is incorporated into local productive processes, and transforms parts of itself into wages and taxes. Its value is increased by the exploitation of local labor forces, which transform nature and produce goods that realize again the life of this capital when staples (oil, copper, bananas, etc.) are sold in the external market.[37]

In contrast:

> The case of contemporary dependent industrializing economies controlled by multinational corporations gives the impression of a return to the enclave type of economy. However, although initial accumulation often results from external investment (though not necessarily, because multinational corporations do often use local funds to invest), there is an important difference vis-à-vis enclave economies: a substantial part of industrial production is sold in the internal market.[38]

Such investment activities in many areas of Latin America have created situations of what is termed "the new dependency." The argument is that industrial capitalism in the context of such investment

> . . . creates concrete situations of dependency with features distinct from those of advanced capitalist societies . . . peripheral industrialization is based on products which in the center are mass consumed, but which are typically luxurious consumption in dependent societies. Industrialization in dependent economies enhances income concentration as it increases sharp differences in productivity without generalizing this trend to the whole of the economy . . . industrialization in the periphery increases disparity of income among wage earners accentuating what has been called in Latin America the "structural heterogeneity."[39]

In addition, such production and consumption patterns mitigate against the development of indigenous capital goods production capabilities. When such factors are viewed in relation to patenting and other practices which tend to inhibit the development of local research and development capabilities, concrete situations of dependency result.

> From the economic point of view a system is dependent when the accumulation and expansion of capital cannot find its essential dynamic component inside the system. In capitalistic economies the crucial component for the drive to expand is the capacity to enlarge the scale of capital. This cannot be done without the creation of new technologies and continuous expansion of the production of "capital goods," that is, machinery and equipment, to permit the continuing growth of enterprise expansion and capital accumulation.[40]

The role of the modern transnational corporation as an institution in such a process is elemental. As will be pointed out below, such institutions have become a major focal point of most participants from peripheral areas who seek to restructure the global political economy.

It is in such a context that we can now explore the interrelationships between the evolving institutional arrangements of the ISA and the principle that exploitation of the Area should benefit all mankind. The common heritage principle was at inception and remains a very ambiguous notion. The basic notion seems to be rooted in the belief that the Area and its resources should not be subject to appropriation and/or control exclusively by a relatively small group of governments and/or private exploiters. Instead, any benefits derived from such exploitation should be enjoyed by all or utilized to help those most in need. Some insights into the principle can be gained by viewing the development of the concept in the context of the call for the establishment of a New International Economic Order (NIEO).

## The NIEO

On May 1, 1974 the Sixth Special Session of the UN General Assembly adopted a Declaration and Program of Action on the Establishment of a New International Economic Order (Resolutions 3201 [S-VI] and 3202 [S-VI]). Following on the heels of the 1973-74 oil embargo and oil price increase, these resolutions represented a direct, yet somewhat tempered, challenge to the structure of the established global social order. The concerns represented in the documents were not new. Indeed, Robert Jordan referred to the NIEO as a "fuse which had been burning for many years."[41]

A decade earlier delegations from peripheral areas had gained a large enough majority to establish a special forum within which to seek more preferential terms of trade vis-à-vis core areas. It was out of these UNCTAD origins that a Third World caucusing mechanism, the Group of 77, emerged. Through UNCTAD and various other multilateral mechanisms, the Group of 77 had refined a series of demands for restructuring the global political economy.

Jordan has classified NIEO concerns into four broad categories: (1) economic sovereignty; (2) trade; (3) aid; and (4) participation.[42] Such a fourfold classification seems to capture very succinctly the various important elements of the NIEO program. A substantial amount of concern has been expressed regarding the lack of real sovereignty (i.e., freedom from outside control and undue outside influence in national political-economic affairs and processes). Jordan has subdivided this concern into four subparts: (1) freedom from intervention; (2) control over natural resources; (3) regulation of transnational corporations; and (4) the right to nationalize foreign holdings. In general, the target of such concerns was the transnational corporation. A new relationship between transnationals and host governments lay at the heart of these concerns.

In the area of trade the major concern was with market structures. On one hand, a means by which to stabilize commodity markets and, thus, export earnings, was desired. Such means was to be created by establishing international regulatory machinery and other institutional arrangements. On the other hand, preferential access on a nonreciprocal basis to industrialized market economies was also sought. The Lomé Conventions between the governments of the EEC and more than 46 African, Caribbean, and Pacific (ACP) governments have attempted to deal with such concerns. The ACP governments were granted preferential access to EEC markets on a nonreciprocal basis. Also, a compensatory finance scheme (STABEX) was created for 12 commodities to help stabilize export earnings of ACP governments. In addition, EEC governments agreed to substantial increases in assistance to these governments both in 1975 and 1979.

Such official development assistance (ODA) is part of the third general area of NIEO concern: aid. The Group of 77 has called for an increase in such assistance on concessional terms. Aid is also requested in the forms of debt rescheduling and the transfer of technology. This latter concern also focuses more generally on the improvement of scientific and technological capabilities in peripheral areas.

Finally, the matter of participation is a very critical one. Jordan asserts that ". . . this issue, more than any other, lies at the heart of the developing states' demands for a new international order."[43] While the first category above focused on sovereignty, the concern with participation focuses more on equality. Such equality not only pertains to decision-making mechanisms, but also with respect to outcomes. The Charter of Economic Rights and Duties of States (UN General Assembly Resolution 3281 [XXIX], Chapter II, Article 10) clearly underscores the concern when it states:

> All states are juridically equal and . . . have the right to participate fully and effectively in the international decision-making process . . . and to share equitably in the benefits resulting therefrom.

Thus, broadly defined, participation looms large in the Group of 77's preferred restructuring of the global political economy.

These issues were raised time and again in various conferences during the 1970s. In addition to the Sixth Special Session of the United Nations General Assembly held in 1974, special sessions in 1975 and 1980 also provided the opportunities to press NIEO demands. The Conference on International Economic Cooperation (CIEC) held from December 1975 to June 1977 allowed special focus to be placed on energy, raw materials, financing, and development concerns. Also, UNCTAD IV and UNCTAD V, which were held in 1976 and 1979 respectively, provided fora within which to work toward an Integrated Programme for Commodities, as well as a more comprehensive set of reforms.

When the Group of 77 presented its demands at the Sixth Special Session in 1974, ideologically charged rhetorical debate ensued. Spero reports that CIEC negotiations remained calm and rational. She goes on to state that "UNCTAD IV . . . clearly demonstrated the new unity and offensive strategy of the South."[44] She then reports that at UNCTAD V major confrontation was also avoided. It appears that after the very emotional "coming out" of the Declaration and Program of Action on the Establishment of a New International Economic Order, at least a degree of willingness to seek solutions in a cooperative fashion has prevailed within the Group of 77. Spero observed that:

> Over a decade of common action in international forums has given the group an understanding of how to bargain jointly with the North. Members now develop their positions early and come well prepared for international negotiations. An important force for unity and strength has been the emergence of OPEC leadership. Some oil producers, Algeria and Venezuela in particular, have been willing and able to mobilize a common Southern front and to link the oil issue with other Third World issues.[45]

## Seabed Exploitation and the NIEO

Just one month after the delegates at the Sixth Special Session of the UN General Assembly adopted the NIEO Declaration and Program of Action, substantive deliberations began in UNCLOS III. However, even before the Sixth Special Session or UNCLOS III, linkages were made between seabed exploitation and Third World concerns, such as commodity price stabilization and transfer of technology.

Some of these linkages were very direct, such as UNCTAD Resolutions 51 (III) and 52 (III) in May of 1972. These resolutions called for a study of the adverse effects of seabed exploitation on fisheries and mineral prices (Resolution 51) and a moratorium on seabed mining (Resolution 52).

Other linkages were somewhat less direct at the time. For example, Edward Miles discusses the cooperative relationship which was emerging during this period between the UN Secretariat and the UNCTAD Secretariat over the preparation of technical reports for the Seabed Committee. He suggested that "these reports began to forge a link between the seabed issue and the question of global commodity policies."[46] It should also be noted that various other topics deliberated in Subcommittee I were inherently linked to NIEO concerns. Especially noteworthy in this regard were the topics of "Who may exploit the area?" and the notion that seabed exploitation was to be for the benefit of mankind as a whole.

Once the conference got underway and substantive discussions in Committee I began in earnest, linkages between seabed exploitation, the NIEO, and UNCTAD became fairly overt. Miles reports that at Caracas Latin American delegations attempted to force a debate on the economic implications of seabed mining in order to provide the ISA with the competence to deal with commodity price stabilization concerns. He goes on to report that representatives of UNCTAD were clearly present within the Group of 77 delegation and made "very forthright recommendations in support of production and price controls . . ."[47]

> Moreover, the role of the UNCTAD reports and representatives in the debates . . . was to raise the possibility of instituting production and price controls as one step on the road to comprehensive commodity price controls. The rationale for this is a potent argument in the Group of 77 and facilitated the submergence of apparent differences in the interests of land-based producers versus consumers in that group.[48]

The leader in attempting to forge such linkages, and thereby mobilize the Group of 77 into a cohesive political force, was the Algerian delegation. Similar to the role played by the Algerian delegation at the Sixth Special Session of the UN General Assembly, and elsewhere at that time, an attempt was made to unify Group of 77 and OPEC support behind a common bargaining position vis-à-vis the Group of Five. Miles argues that a ". . . distinguishing development of the Geneva session was what could fairly be called the triumph of the Algerian initiative within the Group of 77 in Committee I to link the oceans issue with the global commodities problem in the context of claims for a New Economic Order."[49]

This Geneva session in 1975 followed a truly astounding year for members of the Group of 77. Not only had

OPEC oil-producing governments withstood strong opposition to their 1973 oil price increases, but the Group of 77 had pushed through the General Assembly both the NIEO Declaration and Program of Action and the Charter of Economic Rights and Duties of States (UN General Assembly Resolution 3281, December 12, 1974). This latter document had even contained a specific article (Article 29) which dealt explicitly with the seabed exploitation question. Thus, a variety of external factors related to the NIEO served to help set the context for subsequent seabed exploitation regime activities.

However, internal dynamics served to provide a bridge between such external factors and conference diplomacy. Miles describes such dynamics as a "game."

> The "game" . . . involved Algeria as prime mover, but, since Caracas, aided significantly by Tanzania and Mauritania and supported occasionally by China. The point of this "game" was to link the seabed issue tightly with demands for a New Economic Order. Most people in Geneva, and this means most people in the Group of 77, began in fact to play this game and this accounts for the enormous increase in representation in the Group of 77 Contact Group as well as in the Working Group of Committee I. It affected the entire proceedings and threatened the compromises arrived at because "for most countries the issue involves no serious substantive interests . . ." for most people in Committee I, therefore, the seabed was an expendable issue in substantive terms and many delegations who attended the Working Group did not have normal instructions for their governments.[50]

Thus, the seabed issue was linked to demands for a NIEO resulting in a substantial expansion of the seabed exploitation regime. Through such an expansion, group cohesion has been established and maintained.

But the role of NIEO concerns within UNCLOS III in general and Committee I in particular was more than that of a "game." In an unusually frank explication of the nature of the Informal Single Negotiating Text (SNT) drafting procedures, Paul Engo confessed:

> I was compelled in some instances to look outside and beyond the unproductive debates that had dominated the session. . . . I worked in the light of the provisions contained in the Declaration of Principles. . . . Also of considerable importance for me was another international document commanding wider universal support: the Declaration on the Establishment of a New International Economic Order. . . . I could do this because I had a free hand (A/CONF.62/L.16).

The SNT indeed reflected many of the desires and positions of the Group of 77.

Such linkages were made time and again at UNCLOS III. Although a great many of these linkages with NIEO concerns were generated internally to the conference, others were generated externally. The issue linkage between seabed exploitation and global commodity problems within UNCTAD is a case in point. What began as an implicit linkage in the early 1970s, resulting from inter-Secretariat reports, became very explicit by the end of the decade. Resolution 108 (V) adopted by the Trade and Development Board of UNCTAD at its 170th meeting on June 1, 1979, warned that "States which might take . . . unilateral actions have to assume the responsibility for their consequences both with respect to their impact on the United Nations Conference on the Law of the Sea and with regard to the negotiations on commodities related to the exploitation of mineral resources of the seabed . . . ." Although such a resolution is of little consequence for directly influencing such activities, it might serve as an instrument for helping to bolster cohesion among Committee I delegations from the Group of 77.

One of the most interesting aspects of seabed regime social processes has been the evolution of negotiating positions of delegations with respect to various seabed exploitation concerns.[51] It was noted that even though the negotiations proceeded in a somewhat erratic fashion, the positions of conference participants slowly evolved toward a general consensus on a Draft Convention. However, given that such a consensus was not universally shared by all issue regime participants (especially participants from the U.S. hard minerals industry), the overall nature of seabed regime structure and process remained somewhat fluid and subject to the possibility of erratic change.

Within the context of UNCLOS III, though, the general nature of the institutional arrangements, which had evolved from years of negotiation, mitigated against substantial change. As illustrated in Figure 5-1, such arrangements represented somewhat of a "middling solution" between various extreme positions. Although observers from core areas have been quick to point out that such extreme negotiating positions did not represent what they term "real substantive interests" in seabed exploitations, these arrangements did represent a negotiated compromise. As discussed earlier, the seabed institutional arrangements were part of a much larger Law of the Sea "package."

Regardless of answers to the rather normative questions of who benefited or lost more under such arrangements or who gave up more in arriving at such a compromised position, the Draft Convention provisions with respect to "Who may exploit the area" represented a general shift in negotiating positions back toward a more central point. Figure 5-2 compares the position of the Draft Convention with changes which had occurred during the 1971-75 period in delegations' preferred positions with respect to such arrangements. While certainly not in direct accord with the

Figure 5.1. Contending Positions on the Nature of an International Seabed Authority

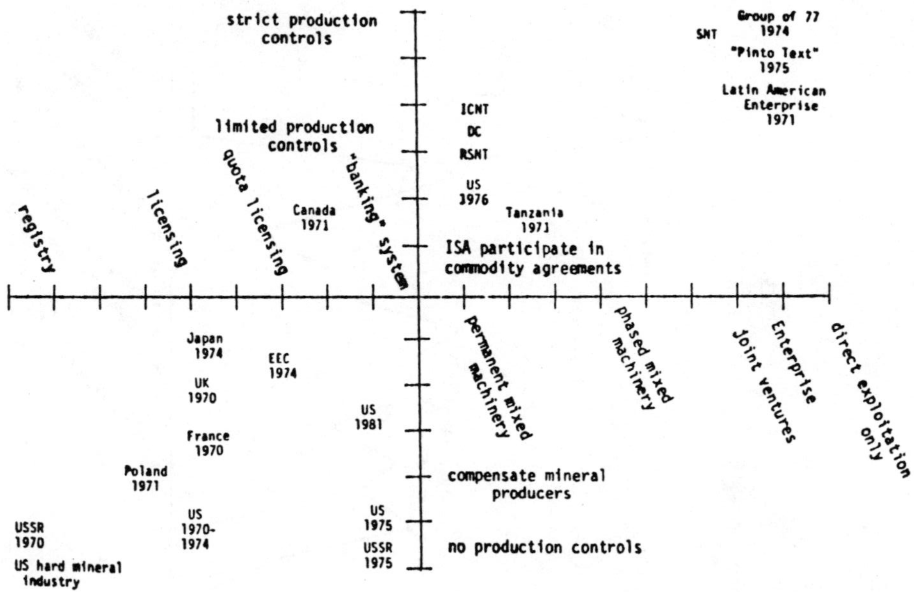


Sources: Friedheim and Durch, "The International Seabed Resources Agency Negotiations and the New International Economic Order," International Organization 31 (Spring 1977):354; Oda (1972-1979); Neptune (various issues); and other formal statements, proposals, and draft treaty texts.

Figure 5.2. Who May Exploit the Area? Trends in Delegate Preferences 1967-1975*

*Data includes both estimate and data-based scores of delegate preferences taken from verbatim and summary records of official policy statements.

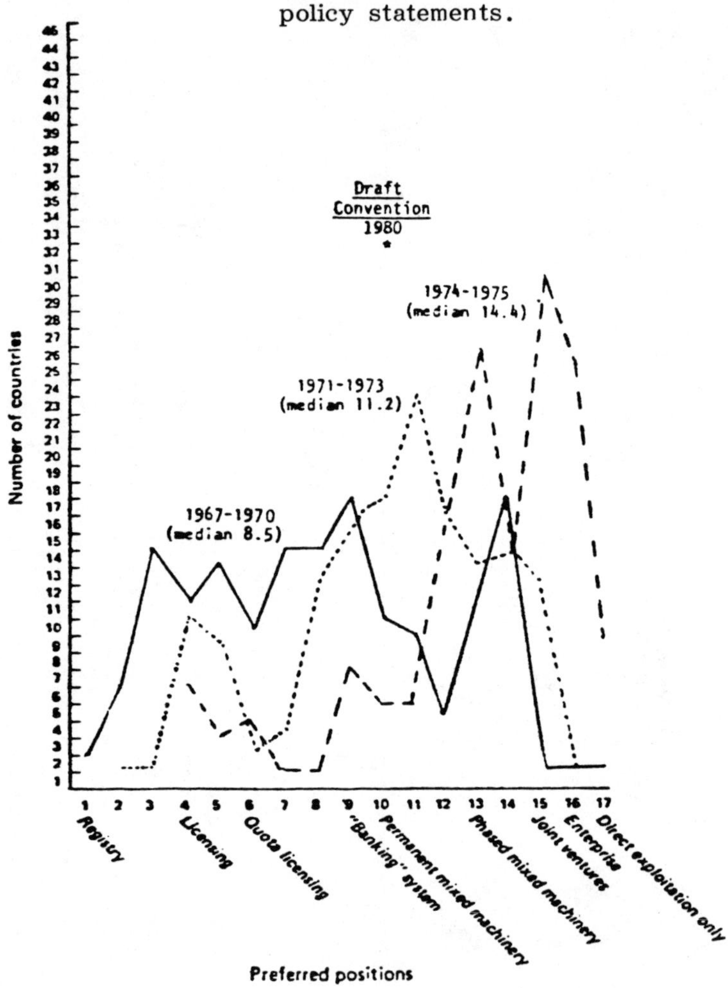


Source: Friedheim and Durch "The International Seabed Resources Agency Negotiations and the New International Economic Order," International Organization 31 (Spring 1977):357-360.

overwhelming preference for a strong Enterprise which existed in 1974-75, the position of the Draft Convention was reflective of various aspects of the larger global political economy.

The year 1974 represented the high point of the NIEO movement as discussed above. During that year participants from the Group of 77 maintained a relatively unified position and put forth a rather bold set of demands. In various negotiations in later years, however, members of the Group were relatively more prone to compromise and to proceed slowly on difficult issues. Such changes were reflected in seabed regime activities.

A great many of the NIEO concerns were accommodated within the delicate compromise positions of UNCLOS III. Much more explicitly and comprehensively than at Bucharest or anywhere else, the basic concerns of the Group of 77 were addressed. The question of sovereignty was creatively dealt with on two levels. First, concerns regarding control over natural resources within national jurisdiction were directly addressed by the establishment of 200-mile exclusive economic zones (EEZs). On the other hand, the notion that the Area be considered the common heritage of mankind became manifest in the provisions regarding the establishment of the ISA. Also, the scope and nature of the ISA addressed the concern over controlling the activities of transnational corporations. The seabed mining activities of such corporate enterprises (as well as state-owned enterprises) would be closely regulated.

Concerns regarding the nature of commodity markets were dealt with very explicitly in Article 151 on "production policies."

> . . . the Authority, acting through existing forums of such new arrangements or agreements as may be appropriate . . . shall take measures necessary to promote the growth, efficiency and stability of markets for those commodities produced from the resources of the Area, at prices remunerative to producers and fair to consumers. . . . The Authority shall have the right to participate in any commodity conference dealing with those commodities and in which all interested parties including both producers and consumers participate. . . .

This article went on to spell out a rather specific production policy by which to achieve such objectives.

Operationalizing the general notion that exploitation of the Area should be for the benefit of all mankind has been a very difficult task. Underlying the difficulties in accomplishing such an objective is basic global inequality. Thus, the NIEO concerns with transfers of technology and financial capital have been central to the task of bringing the concept of the ISA into reality. Under Part XI, Article 144 of the Draft Convention, the ISA is to take measures to acquire technology relating to the activities in the Area and to "promote and encourage" the transfer of such technology to peripheral areas. The "basic conditions of prospecting, exploring and exploiting" the Area (Annex III) specify explicit procedures for technology transfer in Article 5. Similarly, financing for the ISA and Enterprise has been carefully worked out.

The question of participation has also been a critical concern since early Seabed Committee deliberations. Such concerns have centered around the distribution of functions, powers, composition and decision-making procedures within the organs of the ISA. The development in Geneva in 1980 of the three-tiered system of voting for the Council was a major breakthrough in this area. However, such an arrangement represents only one aspect of a very delicate set of compromises regarding participation in the ISA.

As with the other issues discussed above, these provisions were negotiated in the context of a "package" involving a much larger set of Law of the Sea concerns. Also, negotiations within UNCLOS III were conducted in the context of other environmental constraints. Although the activities within the Seabed Committee and UNCLOS III tended to dominate seabed regime activities after 1970, it should be remembered that activities in other institutional arenas were also influencing the evolution of structural changes within the issue regime. The formation and subsequent activities of multinational corporate seabed mining consortia and the threat and subsequent approval of unilateral seabed mining laws within national governments (i.e., the United States, France, and the United Kingdom) were among the most important in these institutional arenas.[52] These various institutional arenas were, in turn, enveloped within the constraints of a much larger global political economy.

## THE ISA AND THE COMMON HERITAGE OF MANKIND: A SPECULATIVE CONCLUSION

The concerns expressed in the call for a New International Economic Order inherently focus on the very essence of the structure of global political economy. The unequal division of labor within the capitalist world economy and the resulting institutional manifestations are seen as major impediments to the goal of fulfilling basic human needs within the peripheral areas. The spirit of the "common heritage of mankind" and the principles and objectives specified in the NIEO Declaration and Program of Action emanate from such a concern. In the seabed exploitation regime, however, little has been done to translate such a "spirit" into meaningful reform.

The global concentration of control over resources has been the primary target of proposals for reform. The "solutions" which have predominated have been research, financial capital and intensive organization. Although very little actual movement has occurred, substantial transfers of financial capital and technology have been proposed. A belief seems to exist that such transfers, when linked with efficient organizational structure, will somehow lead to change within the global political economy, thus alleviating or eliminating structural violence. In this regard, the proposals and activities within the issue regime have been very similar to the Declaration and Program of Action and other NIEO activities and statements.

Beneath the well-worn rhetoric attacking global inequality lies very little in the way of proposals or action for bringing about any real transformation within the global social order. Such a lack of movement, however, is not surprising. Most participants from peripheral areas continue to play the "game of interdependence" within the capitalist world economy. The language as well as the institutions that are utilized are products of such a global order. The United Nations Conference on Trade and Development, UNCLOS III, OPEC, and the Group of 77 all represent manifestations of a global political economy and social order that evolved from a European state system. The evolution of such an interstate social order has been closely associated with the evolution of the capitalist world economy. As discussed earlier, state mechanisms play very instrumental roles within such an order. Negotiations among state mechanisms from the various resource zones will not only be conducted on a very unequal basis, but, more importantly, represent essentially an exercise in system reinforcement. Bargaining among asymmetrical units within such a system cannot produce real structural equality therein.[53] An unequal division of labor is needed for maintenance of the capitalist world order.

From a common heritage perspective the provisions of the Convention related to the transfer of technology (especially Annex III, Article 5) are of critical importance. In this regard, the provisions do provide some degree of discretion to the Enterprise. For example, the Enterprise, not the commercial operator, determines what technology is to be transferred and when it is needed. However, the entire technology transfer arrangement is based on a licensing system. Although such a system of property rights protection might not inherently prohibit meaningful transfers of technology, when applied to a structural relationship involving substantial asymmetry, such licensing arrangements can actually increase dependency, as discussed above. In order to guard against the establishment of such a "technological dependence syndrome," the Enterprise will need a very strong research component.

The notion of transferring seabed mining technology to "developing States" (Convention, Part XI, Article 144) appears on the surface to have far-reaching implications. Perhaps for a few governments and corporations in semi-peripheral areas such a provision might hold promise. However, for the overwhelming number of governments in the periphery and nearly all of the peoples in those areas the provision offers little meaningful assistance. The technology required to mine manganese modules from three miles beneath the ocean's surface is of little utility for providing basic human needs in peripheral areas. Furthermore, to the extent that Third World transnational corporations invest tremendous sums of financial capital in such activities, inequality and other conditions associated with the "new dependency" are likely to intensify.

Equally important for consideration in this regard is that most governmental officials across all zones are so linked into the capitalist world economy that proposals for actual structural transformation within the system appear threatening. Their positions of social control are generally very closely tied to stability within national political economies. As most often is the case in peripheral areas, conditions within most such national economies have become and remain linked to the larger global political economy.

> Peripheral economies, even when they are no longer restricted to the production of raw materials, remain dependent in a very specific form: their capital-goods production sectors are not strong enough to ensure continuous advance of the system, in financial as well as in technological and organizational terms. So, in order to go ahead with economic expansion, a dependent country has to play the "interdependency" game, but in a position similar to the client who approaches a banker. Of course, clients usually develop strategies of independence and can try to use the borrowed money in productive ways. But insofar as there are structural borderlines, successful attempts are not an automatic output of the game.[54]

The "game" in global negotiations then often becomes one of minimizing losses and promoting <u>individual</u> gain. Little of substance with regard to system transformation emerges from intergovernmental mass movements or organizing peak associations, such as the Group of 77.

Other forces within the global political economy also mitigate against maintaining cohesiveness within such groups. The ramifications of the 1973-74 and 1979 oil price increases are a case in point. Although some governmental representatives from peripheral areas (especially Algeria) were able to employ the first major OPEC initiative to help forge a "Third World coalition," the impact of such price increases has had a self-destructing impact on the cohesion of the Group of 77. The borrowing cycles and debt servicing problems which have resulted have tended to drive a wedge between oil-producing and oil-consuming governments.

In addition, such actions have had differential impacts upon Group of 77 governments within the oil-consuming category. As discussed earlier, those relatively few governments (NICs) which might reasonably be said to have attained or were on the verge of attaining semi-peripheral status were in most cases able to borrow and/or otherwise creatively deal successfully with the 1973-74 oil price hikes.

Most peripheral governments were not able to cope adequately. Even though the second major oil price increase in 1979 also seriously affected the NIC governments, again there existed both quantitative and qualitative differences with regard to the nature of the impact on NICs and other peripheral national political economies. The point of this example is not that a cohesive Group of 77 is impossible to attain. Indeed, the tremendous balance of payments surpluses accruing to OPEC governments could go a long way in alleviating debt servicing problems in peripheral areas. However, such massive increases as assistance from OPEC governments do not appear to be likely.[55] The point to be made is that an important commonality upon which the original Group of 77 coalition was based, the condition of underdevelopment and poverty, is rapidly eroding. This point is underscored by the fact illustrated earlier that two of the world's 50 largest financial institutions are based in so-called NICs.

Having made such a point, however, it should be said that various aspects (if not most) of the NIEO program might very well succeed. Outside of the rhetoric there is little of substance which really challenges the basic structure of the capitalist world economy. In fact, OPEC represents a variant of a very classic capitalist manifestation: cartelization. What will not emerge, however, from the NIEO is the elimination of poverty or a major uplifting of human conditions among the masses of peripheral peoples. In all its glory and after all the crimination, the call for the establishment of an NIEO does not really address itself to the basic structural problems underlying such conditions.

What is intriguing, however, is that as a result of activities related to the seabed exploitation regime, some limited movement has been made toward setting the stage for eventual basic needs development. Here, specific reference is being made to Article 56, Paragraph 1 of the <u>Convention</u> that establishes the principle that "coastal states" may claim 200-mile exclusive economic zones (EEZs). According to such a principle, in such a zone a coastal state government possesses:

(a) sovereign rights for the purpose of exploring and exploiting, conserving and managing the natural resources, whether living or non-living, of the seabed and subsoil and the superjacent waters, and with regard to other activities for the economic exploitation and exploration of the zone, such as the production of energy from the water, currents and winds;

(b) jurisdiction as provided for in the relevant provisions of this Convention with regard to:
   (i) the establishment and use of artificial islands, installations and structures;
   (ii) marine scientific research;
   (iii) the protection and preservation of the marine environment;

(c) other rights and duties provided for in this Convention.

By the end of 1976 this principle had become firmly established as a major component, along with freedom of navigation in international straits, of a compromise negotiating "package" in Committee II. As discussed earlier, in Geneva in 1975 the Algerian delegation and others began explicitly to link seabed regime negotiations to such a package. In the years since, the EEZ has become a fairly well-established norm of the Law of the Sea. As of March 1984, over 50 governments, including the United States and the Soviet Union, had proclaimed such EEZs. Although such a principle was not a direct product of negotiations in Committee I, the EEZ can very much be considered a by-product of the seabed exploitation regime as it became institutionalized through the Seabed Committee, which, in turn, evolved into the institutional arrangements of UNCLOS III.

This discussion now can be linked back to the global political economy in general and situations of dependency in particular. The establishment of exclusive economic zones goes further, perhaps, than any other single intergovernmental arrangement or activity since the NIEO Declaration was issued toward the fulfillment of NIEO ultimate objectives. It must be recalled that a basic structural manifestation of situations of dependency is growing inequality within dependent societies. In this regard, Fernando Cardoso and Enzo Faletto contend that "the system of domination reappears as an 'internal' force, through the social practices of local groups and classes which try to enforce foreign interests, not precisely because they are foreign, but because they may coincide with values and interests that these groups pretend are their own."[56] As Cardoso has argued elsewhere, "the local dominated classes suffer a kind of double exploitation."[57] It is precisely the welfare of such peoples that the NIEO claims to serve.

As was discussed above, the Action Program of the NIEO and other related "action" statements really do not address themselves to the kinds of changes which would be necessary to redress such systems of exploitation. Yet the EEZ, in a very important sense, does. Given that 90 percent of known exploitable living and nonliving resources of the oceans fall within such 200-mile limits, these resources in essence also become a "common heritage" of future mankind. The very situation of dependency and the associated condition of underdevelopment, which perpetuate structural inequality and structural violence, also serve to protect and provide somewhat of a "guarantee" against over-exploitation and waste of such resources.

Returning to the main question this chapter addresses, it appears as though a single answer does not suffice. In a general sense the ISA does represent a progressive development of international public Law of the Sea. Such an international institution serves to provide a highly formalized regulatory arrangement for an area (the Area) of ocean space in which heretofore extremely vague notions of res nullius prevailed. Such an institutional arrangement evolved from years of international governmental deliberations within the context of UNCLOS III. However, in a more fundamental sense the institutional arrangements of the ISA represent little direct movement toward the realization of the ultimate objectives of the common heritage principle: exploiting the area for the benefit of all mankind. It seems fair to conclude, though, that the Group of 77 has accomplished much more than ever could have been expected a decade ago, given that these governments have chosen to play within the rules of the system.

## NOTES

1. Draft Statutes were introduced by the delegations from Tanzania (A/AC.138/33), the Soviet Union (A/AC.138/43), Poland (A/AC.138/44), 13 Latin American governments (A/AC.138/49--the "13-Power Proposal"), seven governments from the Disadvantaged Group (A/AC.138/55), Canada (A/AC.138/59), Japan (A/AC.138/63), and Italy (A/AC.138/SC.I/L.24 and A/AC.138/SC.I/L.26).

2. The Vice-Chairman of the U.S. delegation has stated the argument as follows:

> My Government rejects outright the notion that United Nations General Assembly resolutions, including United Nations General Assembly Resolutions 2574D (XXIV) and 2749 (XXV) and irrespective of the majorities by which such resolutions are adopted, are legally binding on any State in the absence of an international agreement that gives effect to such resolutions and that is in force for that State (United Nations Document A/CONF.62/93).

3. Public and private profit-making enterprises in the national economies of Brazil, Chile, China, Gabon, Guyana, Morocco, Peru, Zaire, and Zambia account for a large proportion of the internationally traded copper, nickel, cobalt, and manganese. Foreign exchange revenues generated from the production and marketing of these minerals is vitally important for the functioning of these governments.

4. Edward Miles, "The Structure and Effects of the Decision Process in the Seabed Committee and the Third United Nations Conference on the Law of the Sea," International Organization 31 (Spring 1977):195.

5. Ibid.

6. Ibid., pp. 197-98.

7. Ibid., pp. 207-11.

8. Ibid.

9. The U.S. delegation employed this capability masterfully at the first New York Session in 1976 to "rupture" the cohesiveness of the Group of 77. Ibid., pp. 220-22.

10. Speech by Secretary of State Henry A. Kissinger, "The Law of the Sea: A Test of International Cooperation," New York, April 8, 1976; quoted in Jack N. Barkenbus, Deep Seabed Resources: Politics and Technology (Riverside, NJ: Free Press, 1979), p. 1535.

11. Ibid.

12. The Reagan Administration had decided to pursue a "Reciprocating States Agreement" approach. So while the final UNCLOS III negotiations were in progress, the origins of such an agreement emerged as a result of consultations among governmental representatives of the United States, the United Kingdom, France, and the Federal Republic of Germany.

13. Immanuel Wallerstein, The Modern World System, vols. I and II (New York: Academic Press, 1974, 1980), and The Capitalist World Economy (New York: Cambridge University Press, 1979). Although centered almost exclusively in Europe during this period, Wallerstein refers to the system as a world-system because its very nature was multi-societal.

14. Johan Galtung, "The North/South Debate: Technology, Basic Human Needs and the New International Economic Order," Working Paper 12 (New York: World Order Models Project, 1980), pp. 10-11.

15. Ibid., pp. 10-16.

16. Of course, given the complex nature of contemporary political economy, such exchange generally will not be dyadic. Instead, this relationship will be mediated through a complex system of market structures. Also, it should be noted that only in a very limited sense do laborers have control over their individual or collective contributions to productive processes. The nature of labor's contributions is not determined primarily by laborers but by production managers and others. Thus, the extent to which laborers "share" in the benefits of the capital accumulation process depends in large part on managerial strategies and/or labor's influence within institutions through which market structures and managerial decision making can be affected (including state mechanisms). Returns from productive processes accrue in varying degrees to individuals and groups who can significantly influence and/or control access to various factors of production. Thus, not only do returns accrue to owners of capital and technology, but (to a much lesser extent) to controllers of raw materials and labor as well.

17. Wallerstein, *The Capitalist World Economy*, p. 21.

18. At a very general level Wallerstein differentiates among these areas according to the various factor intensities implied by such a structural division of labor. Core areas are relatively high profit, high technology, high wage, and differentiated production zones. Peripheral areas are characterized by relatively low profits, low technology, low wages, and less diversified production. Semiperipheral areas fall in between (ibid., p. 71).

19. Ibid., pp. 21-22.

20. Ibid., pp. 69-70.

21. Ibid.

22. A major problem with this line of reasoning is that technological progress is seldom constant. Thus, the researcher must pay close attention to the role of technological change in production processes and the resulting implications for change with respect to factor and product markets.

23. In addition, he identifies two other mechanisms that have ensured stability within the capitalist world-economy: (1) concentration of control over means of coercion and destruction in the hands of dominant system-supportive forces; and (2) the pervasiveness of ideological commitment to the perpetuation of the capitalist world-economy (Wallerstein, *The Capitalist World Economy*, p. 22).

24. United Nations, "Multinational Corporations in World Development," in *Transnational Corporations and World Order*, ed. George Modelski (San Francisco: Freeman, 1979), p. 17. "At least 80 percent of the United States affiliates and 75 percent of United Kingdom affiliates are either wholly owned or majority-controlled. In terms of stock of investment, these two countries have placed about 90 percent in affiliates which are at least majority-controlled" (p. 23).

25. Ibid., p. 27.

26. Joan Edelman Spero, *The Politics of International Economic Relations*, 2d ed. (New York: St. Martin's Press, 1981), p. 222.

27. United Nations, "Multinational Corporations," p. 27.

28. Ibid.

29. Spero, *International Economic Relations*, p. 223.

30. Ibid., p. 224.

31. Ibid.

32. Ibid., pp. 204-05.

33. Constantine Vaitsos, "Patents Revisited: Their Function in Developing Countries," *The Journal of Development Studies* 9 (October 1972):71-97.

34. Lynn K. Mytelka, "Technological Dependence in the Andean Group," *International Organization* 32 (Winter 1978):101-39.

35. Ibid., p. 134.

36. Ibid., p. 102.

37. Fernando Cardoso and Enzo Faletto, *Dependency and Development in Latin America* (Berkeley: University of California Press, 1979), pp. xviii-xix.

38. Ibid., p. xix.

39. Ibid., p. xxii.

40. Ibid., p. xx.

41. Robert S. Jordan, "Why the NIEO? The View from the Third World," in *The Emerging International Economic Order: Dynamic Processes, Constraints and Opportunities*, ed. Harold Jacobson and Dusan Sidjanski (Beverly Hills: Sage Publications, 1982), p. 8.

42. Ibid., pp. 12-23.

43. Ibid., pp. 22.

44. Spero, *International Economic Relations*, p. 209.

45. Ibid., p. 204.

46. Miles, "Decision Process," p. 179.

47. Ibid., p. 193.

48. Ibid.

49. Ibid., pp. 197-98.

50. Ibid., p. 207.

51. For good detailed discussions and analyses of such changes, see Robert L. Friedheim and William J. Durch, "The International Seabed Resources Agency Negotiations and the New International Economic Order," *International Organization* 31 (Spring 1977):342-84; and Barkenbus, *Deep Seabed Resources*.

52. For a discussion of such unilateral seabed mining initiatives, see _Neptune_ 18 (August 1981):1.

53. Such a statement does _not_ mean to imply that international governmental negotiations and cooperation are useless mechanisms for bringing about better living conditions for the overwhelming majority of the world's peoples. However, structural constraints mitigating against the attainment of such an objective need to be clearly identified and recognized.

54. Cardoso and Faletto, _Dependency and Development_, p. xx.

55. Like most other official development assistance, OPEC governmental development assistance declined markedly in real terms during the last half of the 1970s.

56. Cardoso and Faletto, _Dependency and Development_ p. xvi.

57. Fernando Cardoso, "The Consumption of Dependency Theory in the United States," _Latin American Research Review_ 12 (1977):13.

# 6

# UNITED STATES POLICY AND THE MANAGEMENT OF MARINE MINERALS

*Christopher Herrick*
*Robert S. Jordan*

## INTRODUCTION

The role of the international community in the management of ocean space has grown dramatically in the past quarter century.[1] The principal area of concern has been the area of the high seas. However, international negotiations and agreements have also had an increasing effect upon countries' management of their coastal waters.

This chapter examines briefly the history of international agreements and conference activity with particular reference to the management of the territorial sea and surrounding waters prior to the Third United Nations Conference on the Law of the Sea (UNCLOS III). It next describes the provisions of the Law of the Sea Treaty and analyzes the possible impact of many of these provisions on the management of U.S. coastal waters. The chapter then briefly explores the impact of other international agreements on the management of U.S. coastal waters, before focusing on the important policy issue of technology transfer in multinational negotiations. We conclude with a discussion of the role of international organizations in shaping ocean policy.

## HISTORY OF INTERNATIONAL CONCERN FOR THE OCEANS

International agreements and conference activity concerning the oceans may be divided into three periods. Prior to 1958, there was little organized effort toward the development of an international regime for the oceans. Rather, the regime which emerged was the result of centuries of discrete actions incorporated as customary law by individual nations or small groups of nations. Conference activity specifically directed toward the development of a more complex international regime for the oceans may be said to have begun in 1958, but the participation in this United Nations Conference on the Law of the Sea, was generally confined to the developed countries. The conventions which issued from UNCLOS I did not serve as the basis for the development of a coherent, stable, or comprehensive regime.

By the mid-1960s, man's ability to exploit a wider range of marine resources over a considerably greater area of the oceans was apparent. At the same time, the now independent Third World nations began to express an increased desire to exert greater control over the resources lying off their shores. In addition, the developing states expressed increased interest in participating in the management of ocean resources, and, if possible, in the exploitation of a wider range of marine resources. These factors coupled to provide the impetus for the convening of a more universal international conference for the purpose of considering a broad array of issues concerned with the management of the oceans.

Until the twentieth century, international concern for the oceans centered on their use as a source of food, as a medium of commerce, and to a lesser extent, as a medium by which to project national power through military means. The international regime for the oceans, with supporting laws and conventions, reflected these purposes and remained relatively stable for a considerable period. The regime was directed toward protection of near-shore areas, while the seas beyond remained virtually unregulated. On the matter of the extent of the near-shore area subject to national jurisdiction, there was general, though not complete, agreement.

While the three-mile territorial sea enjoyed general recognition as the major delimiting boundary of near-shore[2] regulatory activities, it was not the only zone in which the coastal state could exert its authority. Even the United States, a strong supporter of the concept of a three-mile territorial sea as the limit of a nation's seaward jurisdiction, has, since the earliest days of the Republic, advocated more extensive zones for specific purposes. Among the extended zones advocated or adopted by the United States were: (1) a neutrality zone reaching to the Gulf Stream (advocated by Jefferson); (2) a customs zone of four leagues declared in 1799; (3) extension of the zone in which prohibition laws were enforced to a distance of four leagues; (4) the Liquor Treaty of 1924 which extended the enforcement zone to a maximum of approximately 30 miles; and, (5) the security zone of the Panama Declaration which stretched seaward a distance of several hundred miles.

International conference activity prior to 1958 was chiefly directed toward refining or codifying this system. For example, the Hague Codification Conference of 1930, under League of Nations sponsorship, devoted considerable attention to developing more precise definitions of the width of the territorial sea, of the baseline to be used in determining this limit, and of the rights of the coastal state to manage activities within this area.[3] This conference also directed attention toward the prospect of instituting a contiguous zone beyond the territorial sea. In addition, the conference anticipated the actions of individual states, such as the United States, in drawing attention to the possible need for, and problems related to, coastal state control of the continental shelf.

Although the 1930 Hague Conference broached the subject of expanded national claims to jurisdiction over ocean space, the new era in ocean politics may be said to have begun in 1945. In that year the United States did, in fact, unilaterally declare its jurisdiction over the resources of the vaguely defined continental shelf and certain fisheries of the waters off its shores. In the years immediately following this action, a number of other countries declared similar vaguely defined conservation or resource management zones.[4] Reacting to the Truman Proclamations and similar declarations by other countries, the United Nations International Law Commission initiated a comprehensive review of existing ocean management problems and practice. Among the issues considered were: the breadth of the territorial sea; the right of jurisdiction over the continental shelf; fisheries management; conservation of marine resources; and the definition and regulation of activities on the high seas.[5] The release of the commission's final report in 1956 provided impetus for the United Nations to convene an international conference for the purposes of standardizing and codifying international practice.

UNCLOS I was convened in 1958 at Geneva. After lengthy deliberations, the 86 countries participating in the negotiations were able to develop four conventions which were subsequently ratified by the United Nations General Assembly. These included:

(1) A Convention on the Territorial Sea and Contiguous Zone;
(2) A Convention on the Continental Shelf;
(3) A Convention on the High Seas; and
(4) A Convention on Fishing and the Conservation of the Living Resources of the High Seas.

Of the four conventions, the first two are of particular interest.[6]

The Convention on the Territorial Sea and Contiguous Zone was notable for its lack of definition. While it did specify the low water line as the landward baseline from which to calculate the width of the territorial sea, it did not specify any figure for the seaward limit of this zone. Further, although the convention delimited the outer border of the Contiguous Zone at 12 miles from the low water line, it did not specify the inner boundary of the zone.

The Convention on the Continental Shelf was equally vague in its definition of the continental shelf area. Article 1 of the convention defined the continental shelf area:

> as referring (a) to the seabed and subsoil of the submarine areas adjacent to the coast but outside the area of the territorial sea, to a depth of 200 meters, or beyond this limit to where the depth of the super-adjacent waters admits to the exploitation of the natural resources of said area . . . .

The Second Law of the Sea Conference (UNCLOS II), convened in 1960, again considered the problem of defining the limits of the territorial sea and the continental shelf. This conference, however, failed to arrive at any consensus regarding either issue. As a result, the conventions drafted at the earlier conference which passed into effect in 1964 have remained in effect to the present time.

The deficiencies of the above two conventions, however, became apparent soon after their entrance into force. In failing to delimit the territorial sea, the conventions left states free to extend their territorial seas to virtually any width. A few states did, in fact, declare territorial seas stretching hundreds of miles seaward. Further, in employing the 100 fathom (200 meter) depth as the delimiting measure for the continental shelf, the convention made it possible for states with broad continental shelves to claim areas hundreds of miles to sea while other states with narrower shelves were limited to claims of only a few miles. Of perhaps greater importance were the difficulties which quickly became apparent with the definition presented in the final clause of the article. A literal reading of, "to a depth where the super-adjacent waters admits of the exploitation of the natural resources of said area," would suggest that any state could claim as far seaward as its technology would allow it to develop. Thus, a country such as the United States, with a high level of technology, could conceivably claim substantial portions of the ocean floor for itself under this definition.[7]

Within three years of the time that the 1958 conventions came into force, it was evident that a number of factors were prompting states to make claims to increasingly broad sectors of ocean space. Oil companies had demonstrated their ability to drill successfully in increasingly deep waters. Other industrial groups were actively exploring the possibility of exploiting manganese nodule deposits on the deep seabed. It was also becoming increasingly evident that unregulated exploitation of coastal fish stocks was endangering the continued productivity of these stocks. Finally, growing concern about the effects of various sources of pollution on the continued productivity of the marine environment prompted states to exert jurisdiction over bordering sections of ocean space.

Although in many cases they had themselves been active in asserting jurisdictional authority over increasingly broad areas of ocean space, those maritime nations relying on unimpeded access to, and use of, wide sectors of ocean space for purposes of commerce and defense were clearly alarmed by the prospects of expanded national jurisdiction over large sectors of heretofore open ocean. On the other hand, developing countries feared that the developed world could and would claim and exploit the resources within and beneath large sectors of the open ocean. This fear was based to some extent on the perception that if developed countries were to exploit seabed resources, they would become less dependent upon raw materials exported by developing states. It was also based on the perception that immediate exploitation of open ocean resources by developed states would leave developing countries few ocean resources to exploit when, and if, they had the technology to do so. In part due to these fears, developed and developing states agreed to convene a third Law of the Sea Conference (UNCLOS III) for the purpose of developing a new regime for the management of the world's ocean space.

UNCLOS III, which convened in December 1973 with 140 nations present, was the largest international conference ever assembled. Moreover, the conference agenda, directed toward the drafting of a comprehensive treaty which could serve as the basis for a new international regime for the oceans, was among the most exhaustive ever attempted in an international negotiation. Nevertheless, the majority of the delegates were of the opinion that these negotiations could be satisfactorily completed within a relatively short period of time. However, despite this initial optimism, delegates were unable to reach final agreement on a number of the issues before the conference, even after virtually ten years of negotiations and six formal drafts of a treaty.

## AN EXAMINATION OF THE TREATY

An analysis of the complex process of negotiations at the multiple sessions of UNCLOS III is clearly beyond the scope of this chapter.[8] So too is an exhaustive analysis of all 320 articles and multiple annexes of the Law of the Sea Treaty. Rather, we will present a brief account of the major divisions of the treaty and a more detailed description of those articles directly applicable to the management of the territorial sea and surrounding waters.

The final draft of the Law of the Sea Treaty was published on October 7, 1982 (UN Doc. A1 CONF. 621122). Of the 320 articles in the Treaty, 21 related to general procedural provisions or final clauses. Another 32 articles concerned the definition of the territorial sea and contiguous zone, or the rights of coastal states to regulate activities within these areas. An additional 31 articles defined the extent of the Exclusive Economic Zone (EEZ) or the continental shelf, and the rights of coastal states to manage activities within these areas. A further 12 articles concerned the issue of straits used for international navigation. Part IV of the Treaty, containing 9 articles, addressed the issue of archipelagic states. The 35 articles contained in Part VII of the Treaty outlined the rights and duties of states, or their ships, on the high seas. The

12 articles of Parts VIII, IX, and X dealt with special geographic situations, including islands, enclosed and semi-enclosed seas, and landlocked states.

With 59 articles, Part XI was the most extensive division of the Treaty. This part detailed the principles governing the area beyond national jurisdiction (the Area), the conduct of activities within the Area, and the development of the resources of the Area. In addition, Part XI described the powers of the international authorities established to manage and exploit the resources of the Area. Annexes III through VIII included detailed provisions concerning the management of the Area.

The protection and preservation of the marine environment also received considerable attention in the Treaty. A number of articles dealt with technical assistance and global cooperation in these areas. Other sections dealt with enforcement of anti-pollution measures or regulations.

Part XIII of the Treaty addressed the issue of marine scientific research. Several articles in this part outlined provisions for marine research within the territorial sea and the EEZ. Other articles established conditions for the conduct of marine scientific research on the high seas.

Part XIV of the Treaty dealt with the development and transfer of marine technology. This section included articles describing ways and means of international cooperation. It also contained an article dealing with the protection of legitimate proprietary interests in technology.

The final substantive division of the Treaty, Part XV, addressed the issue of the settlement of disputes. Several articles described compulsory dispute settlement procedures to be applied in connection with activities within the Area.

Upon assuming office, the Reagan Administration conducted a lengthy review of the Treaty provisions and ultimately concluded that it could not sign the Treaty without substantial revisions. The majority of participating states disagreed. Ultimately, the Treaty was adopted with only minor revisions over the objections of the United States.

## Articles Applicable to the Management of Coastal Waters

Over 40 of the articles of the Treaty were directly or indirectly applicable to the management of coastal waters. The majority of these articles dealt with the territorial sea or the EEZ. Other articles dealt with pollution control and marine scientific research within coastal waters. Finally, the provisions governing activities within the Area may indirectly affect the conduct of activities within coastal waters.

### The Territorial Sea and Contiguous Zone

The articles contained in Part II of the Treaty related to the territorial sea and contiguous zone. These articles were grouped into four sections. The first section provided a general definition of the legal status of the territorial sea. In particular, the article recognized the sovereignty of the coastal state over the area designated as the territorial sea, and also recognized coastal state sovereignty over the air space above and the seabed and subsoil beneath the territorial sea.

The second section of Part II defined the limits of the territorial sea. Article 3 established the right of states to establish "the breadth of this territorial sea up to a limit not exceeding 12 miles." Article 4 set the outer limit of every point of the territorial sea at an equal distance from the baseline. Thus, a state may not declare a three-mile territorial sea along one section of coast and a 12-mile territorial sea along another section. Article 5 established the low water mark as the normal baseline from which to measure the breadth of the territorial sea.

Articles 6 through 13 described special provisions for the delineation of the territorial sea. Each of these could be significant in providing for the potential expansion of the claims of individual coastal states within the United States. Article 6 set the baseline for islands having reefs at the low tide mark of the reef. Articles 7 and 9 provided a series of instances in which it is possible to draw a straight baseline across indentations, the mouths of rivers and unstable deltas on a coast. Article 8 defined the internal waters of a state. Article 10 contained a lengthy series of provisions whereby it may be possible for a state to draw a straight baseline across the mouth of a bay. In particular, this article allowed the drawing of a closing line if the bay is not more than 24 miles from its mouth to its throat, or if it is an "historic bay." Although Article 11 stated that countries may use permanent harbor works as the baseline point, it specifically denied a state the right to designate artificial islands or off-shore installations as the baseline from which they will measure their territorial sea. Article 12 provided that "roadsteads which are normally used for loading, unloading, and anchoring ships, and which would otherwise be situated wholly or partly outside the outer limit of the territorial sea, are included in the territorial sea." Finally, Article 13 provided that low tide elevations wholly situated beyond the breadth of the declared territorial sea from the mainland or an island may not have territorial seas of their own.

Section 3 of Part II addressed the issue of innocent passage within the territorial sea. Article 17 asserted the right of innocent passage through the territorial sea, while Articles 18, 19, and 20 defined the meaning or conditions of innocent passage. Of the remaining articles in this section, three are of particular significance for the management of the territorial sea. Article 21 set forth the rules and regulations which a state may impose upon ships passing through its waters. In particular, Paragraph 1 provided for regulations pertaining to safety, the preservation or conservation of living marine resources, pollution, and marine scientific research, as well as the traditionally accepted right to regulate customs, immigration, etc. Article 22 specifically recognized the right of coastal states to establish sea lanes and traffic separation schemes within the territorial sea. Finally, Paragraph 3 of Article 23 provided that a state may declare temporary security zones surrounding areas of the territorial sea in which activities such as weapons tests are being conducted.

Article 33 defined the area of the Contiguous Zone and the right of the coastal state to regulate activities within that zone. This article provided that the maximum width of the zone is 24 miles from the baseline. In general, the rights granted to the coastal state within this zone are similar to those granted under the 1958 Convention.

## The Exclusive Economic Zone

Part V of the Treaty dealt with the EEZ. Among the articles in this part, 11 are of particular importance. Article 56 outlined the general rights and duties of the coastal state within the EEZ. According to this article, the coastal state is recognized as having sovereign rights over the natural resources of the zone, including living and nonliving resources within and beneath the area. In addition, the coastal state is considered to have jurisdiction over other activities, "for the economic exploitation of the zone, such as the production of energy from the waters, currents, and winds," within the area. Article 57 established the width of the EEZ at 200 nautical miles from the baseline for the territorial sea. Article 57 described the general rights of foreign states within the EEZ of another country. These include the rights of innocent passage and overflight, as well as the right to lay cables and pipelines as long as these comply with the generally applied safety regulations of the coastal state.

Articles 60 to 67 concerned the rights of the coastal state to regulate specific resources within the EEZ. Article 60 confirmed the right of the coastal state to establish and regulate artificial platforms or structures within the zone. Articles 61 and 62 related to the right of the coastal state to manage living marine resources entirely resident within the EEZ. In conformity with existing international conventions, Articles 66 and 67 assigned responsibility for the management of anadromous and catadromous species to the coastal state within whose waters the species spawn or are generally resident. Article 63 dealt with stocks which overlap the EEZs of two or more nations. Apart from suggesting appropriate bilateral or multilateral consultation on management of these species, the article does not prescribe either management criteria or format. Finally, Article 65 affirmed the right of coastal states to act alone or in concert with other interested parties or the appropriate international organizations to limit the taking of marine mammals.

## The Continental Shelf

Articles pertaining to the rights and duties of nations with respect to the continental shelf were contained in Part VI of the Treaty. Article 76 provided a complex definition of the limits of the continental shelf. This definition is based upon either geological factors or a distance in nautical miles. Thus, in simplified terms, the legal limit of the continental shelf may be 200 miles from shore, even if the actual geological configuration is narrower. Conversely, the legally defined continental shelf is limited to a maximum distance of 350 miles from the low water mark or 100 miles distance from the 200 meter isobath, even if the geologically defined shelf is broader. Article 77 established the right of the coastal state to manage the living and nonliving resources on or beneath the bed of the continental shelf. Article 81 established the right of the coastal state to control drilling along the continental shelf.

Articles 77 and 79 enumerated the rights of foreign countries within the continental shelf areas controlled by another nation. These include the right of navigation and overflight of the area. In addition the article recognized the right of all states to lay submarine cables and pipelines along the continental shelf as long as these comply with the general safety and environmental regulations established by the coastal state.

## Articles Concerned with the Regulation of Marine Pollution

Six of the articles contained in the Treaty may be directly applicable to the right of coastal states to institute anti-pollution regulations respecting their territorial seas or EEZs. Article 207 specifically recognized the right of coastal states to establish laws concerning land-based pollution sources. Article 208 ascribed to coastal states the right to implement anti-pollution regulations with respect to activities on the seabed area within their jurisdiction. Paragraph 5 of Article 210 outlined the right of coastal states to establish regulations regarding dumping within their territorial sea or EEZ. Paragraph 3 of Article 211 enumerated the right of coastal states to enact regulations concerning pollution from, and safety standards of, vessels calling at their ports. Finally, Articles 218 and 220 of the Treaty described the right of coastal states to enforce anti-pollution regulations within their waters.

## Articles Concerned with Marine Scientific Research

Articles 245 and 246 outlined the right of coastal states to control marine scientific research within their territorial sea, their EEZ, and their continental shelf area. Within the territorial sea, coastal states are recognized as having an unencumbered right to regulate, authorize, or conduct marine scientific research. Within their EEZs, however, coastal states are recognized as enjoying a more limited right of control. Article 246 directed that coastal states <u>shall</u> grant permission for such research except under previously specified conditions and on a nondiscriminatory basis.

## EFFECT OF AN LOS TREATY ON THE U.S. OCEAN MANAGEMENT REGIME

Many of the provisions contained in the Law of the Sea Treaty would directly affect the management of U.S. coastal waters. Other provisions would indirectly affect the management of the territorial sea and surrounding waters in that they provide international sanction for past unilateral actions by the United States. Still other provisions of the Treaty might indirectly affect the management of U.S. coastal waters in that they may tend to create a regime for the area beyond the continental shelf which private developers would perceive to be hostile to their interests.

### Direct Effects

Those provisions of the LOS Treaty establishing the limits of the territorial sea and EEZ, and the rights of coastal states to manage resources within those zones won general acceptance in the early stages of the UNCLOS III negotiations. The majority of coastal nations have declared a 12-mile territorial sea.[9] As a result this provision has increasingly taken on the status of customary law. Following the declaration of a 200-mile fisheries zone by the United States in 1976, other states have adopted similar resource management legislation. Consequently, the concept of a 200-mile resources management zone embodied in the Treaty has also taken on the status of customary international law. Given the general acceptance of a 12-mile territorial sea and a 200-mile EEZ, the United States would have most certainly come under increasing domestic and international pressure to align the basic framework of this ocean management regime with that of the rest of the world. Should the United States have acceded to a Treaty containing these provisions, the impetus toward declaration of a 12-mile territorial sea and a generalized EEZ would have been even stronger.

Thus, many of the provisions of the Treaty have the effect of clarifying or giving post facto international legitimacy to a variety of resource management legislation enacted on a unilateral basis by the United States in the course of the last 30 years. While granting added international legitimacy to the Outer Continental Shelf (OCS) Lands Act, Article 76 of the Treaty also finally establishes a fixed boundary to the heretofore ill-defined outer continental shelf land area referred to in the OCS Lands Act. Taken as a group, Articles 57, 61, and 62 confirmed the right of the United States to establish the Fisheries Conservation

Zone. Further, Articles 12 and 60 confirmed the right of the United States to establish and regulate traffic about deepwater ports. The provisions of Articles 210, 211, 218, and 220 recognized the international legitimacy of United States Ocean Dumping Legislation and the right of the United States to regulate discharge levels from ships visiting its ports. Finally, Article 65 granted international confirmation to the right of the United States both alone and in concert with other states to enact regulations for the protection of marine mammals beyond its zone of exclusive jurisdiction.

## Indirect Effects

The principal indirect effects relate to the propensity of U.S. private corporations to engage in resource development activities within and beyond the continental shelf area. According to some of the more influential private and governmental leaders, the Treaty provisions dealing with the exploitation of resources in the area beyond national jurisdiction (the Area) are decidedly hostile to private development efforts. In particular, these individuals point to the fact that the provisions of Part V and Annex III would have created major problems for the potential private developer of the resources (especially, though not exclusively, the hard mineral resources) of the Area. In addition, they view the provisions of the Treaty pertaining to representation on, and the powers of, the policy-setting bodies of the international regime (the Council and the Assembly) as being too vague.

Among the provisions of Annex III to which private developers objected are those contained in Article 5, subsection 3e. This subsection would have required the developer and its subcontractors to make their technology available not only to the Enterprise, but also to individual developing states. In the view of many potential private developers, this provision would not have allowed them to properly guard their proprietary rights to the technology which they have developed at great expense. Even though the most recent set of financial terms of contracts presented in Article 13 of the Annex are considerably more reasonable than those contained in past drafts, industry spokesmen have continued to indicate that even these revised terms would prove too onerous to allow them to undertake activities in the seabed area.

Representatives of a number of developed states were skeptical of the feasibility of the provisions relating to the powers and constitution of the ultimate governing bodies for the Area. Generally, they were concerned that membership on the Council could be so maneuvered that states heavily involved in, and in the future dependent on, the development of seabed resources would be unable to protect their vital interests. In particular, these developed states have feared that through domination of the Council, the developing states could so alter existing provisions for the conduct of activities in the Area as to make private exploitation of the resources of the Area technically or financially prohibitive.

Taken as a group, therefore, the provisions of the Law of the Sea Treaty concerned with the management and exploitation of the resources of the Area were seen to go against the interests of potential private developers and the industrialized states which would sponsor their efforts. The extent to which these provisions would prove to be sufficiently onerous to private developers to preclude their participation in the development of the resources of the Area, if the Treaty were ratified by the United States, is open to question. Some industry spokesmen assert that the present provisions would have stifled private interest in developing the resources of the Area, whereas other industry representatives have conceded privately that they could operate within the current provisions, albeit unwillingly and on a reduced scale. However, if the Treaty provisions should have in fact proven too burdensome for private development of the resources of the Area, there might have been as a result increased interest among these groups in more extensive exploitation of near-shore resources.

There is some evidence that the Treaty provisions regarding marine minerals may in fact have proven to be both too onerous for the private mining consortia to contemplate mining under them. As of the spring of 1984 none of the mining consortia have indicated a willingness to file for sites before the December 1984 deadline for initial claims. In addition, the Treaty provisions may have prompted some interest, on the part of the U.S. government at least, toward promoting the exploitation of minerals within the EEZ. The U.S. Department of Interior has recently made available a series of hard mineral lease tracts which have as yet to spark interest on the part of private developers.

## EFFECT OF PROVISIONS IN THE ABSENCE OF A TREATY SIGNED AND RATIFIED BY THE UNITED STATES

The evolution of U.S. policy toward the management of marine minerals was to a significant degree a product of shifting key policy personnel; this was reflected in both perceptions of and sympathy regarding the relative priority to be given to four policy considerations. These four considerations were: (1) the desirability of achieving the precedent of a workable international ocean agreement; (2) the necessity of securing sea lanes and limiting national encroachment on the high seas; (3) the undesirability of establishing a precedent of international control of resources with its consequent potential limitations on U.S. business activities; and, (4) the necessity of securing access to certain mineral resources deemed vital to U.S. industry. Although industry groups directly concerned with mining, together with their congressional allies and certain sectors of the administration, lobbied for the fourth consideration to be given high priority within or without a

Law of the Sea negotiation, the majority of administration departments favored emphasis on the first two considerations. The Defense Department's vital concern with preventing nations from extending their seaward jurisdiction stemmed in large part from the perceived impact of the extended jurisdiction on the United States' ability to project military power around the globe. In the absence of restrictions, certain key straits, as well as vast portions of the hiding grounds of nuclear submarines, could have been closed to the United States.[10] At the same time, the responsible State Department personnel saw the Law of the Sea Conference as representing the most significant chance in years for demonstrating the worth of international cooperation as a means of solving somewhat technical issues.

As mentioned earlier, when UNCLOS III convened initially at Caracas in 1974, the majority of the representatives were of the opinion that a satisfactory Treaty could be drafted in a relatively short time. The majority of countries were also of the opinion that a comprehensive Treaty was essential if the international ocean regime was to be stabilized. As the shape of LOS provisions concerning the marine mineral issue became clearer, however, certain parts of the U.S. bureaucracy, including the Commerce and Treasury Departments, began to express reservations regarding their potential detrimental impact on the prospects for private development not only in the area of seabed mining but, through their precedential nature, on a variety of other potential areas of multilateral negotiations as well. Nevertheless, throughout the ensuing years of negotiations, despite the relatively slow progress and wide divergence of opinion on critical issues at UNCLOS III, delegates from developing and developed nations, including the United States during the Nixon, Ford, and Carter Administrations, continued to assert that a comprehensive Law of the Sea Treaty was both necessary and possible. As the negotiations drew to a close, and the remaining points of disagreement among nations over Treaty provisions became sharper, frustration within and among governments grew. As this occurred, so too did the sentiment in the United States that possibly no treaty was better than a bad treaty ratified, and infinitely better than a bad treaty initialed but not ratified.

Within the United States, dissatisfaction with the provisions of the draft LOS Treaty and the prospects for altering these provisions centered on two points. First, as discussed previously, from the early stages of the negotiations, private groups and their legislative supporters had been interested in ensuring that U.S. corporations would have access to the resources of the deep seabed on reasonable terms and conditions. These interests were concerned with both the content and progress of negotiations at UNCLOS III. As the negotiations dragged on, these groups expressed their concern that U.S. firms would lose their technological advantage if they were not allowed to engage in, or at least plan for, large-scale development in the relatively near future. The seeming inability of U.S.

negotiators to win more favorable terms in the draft Treaty also caused these groups to be concerned that the ultimate LOS Treaty would be decidedly detrimental to their efforts to exploit seabed resources.

Second, other public and private groups expressed concern over the content of Treaty provisions to which the United States had already given tentative approval. This concern was prompted in part by a belief that issues which were thought to be vital during the early stages of the negotiations would no longer be quite so vital. For example, while the right of transit through international straits was still a very desirable provision, its inclusion in a general Treaty was viewed as somewhat less vital than it was at the outset of the negotiations. This was due in part to the perception that it would be possible to negotiate bilateral treaties for the use of the most vital of these straits, which the United States had considered doing. It is also due to the fact that some of the most modern U.S. warships are either physically incapable of transiting these straits or are incapable of transiting the straits in a secure manner. Further, more modern submarines, such as the Trident, no longer have the need of transiting these straits in order to reach assigned stations. Concern over the Treaty was also prompted by fears that some of its provisions would turn out in practice to be double-edged swords. That is, their application in one context would be advantageous to the United States, while their application by other countries would put the United States at a serious disadvantage. This is true of some of the provisions for navigation, pollution control, and scientific research in coastal waters.

Thus, growing dissatisfaction among executive and legislative branch officials with progress in the negotiations became more evident after the Reagan Administration took office in January 1981.

Unlike past administrations, the Reagan Administration was more in sympathy with the arguments regarding both the need to preserve unrestricted access to strategic seabed minerals in the face of uncertain land-based sources and the need to avoid the potentially detrimental precedent of an international management regime not dominated by the United States and the developed states. Moreover, Reagan was willing to replace the existing U.S. Law of the Sea negotiating team with new persons who shared his concern, and in some cases were more adamant regarding the need to avoid the possibly damaging precedents present in the Treaty. For example, at the time that President Reagan replaced the majority of the U.S. delegation to UNCLOS III, the administration issued a statement to the effect that the United States would have to reconsider at length its interests in a Law of the Sea Treaty. In view of the apparent dissatisfaction of the administration and elements of Congress with the state of the negotiations and the product of past negotiations at UNCLOS III, the United States decided neither to sign nor to ratify the Treaty. This raised the question of the impact of the UNCLOS III negotiations on the management of U.S. coastal waters in the absence of a ratified Treaty.

However, certain provisions contained in the Treaty had, in the course of the negotiations, taken on the status of customary law. Among the most significant of these provisions from the perspective of this chapter are the provisions establishing 12 miles as the acceptable breadth of a territorial sea, and the general acceptance of the concept of a 200-mile EEZ. Thus, even in the absence of a unanimously approved Law of the Sea Treaty, there was a strong impetus for the United States to join the majority of states in declaring a 12-mile territorial sea, if only to have a more secure position in dismissing the outlandish claims of some other states to excessively broad territorial seas.

### Impact of Other Negotiations on the Management of U.S. Waters

The Conference on the Law of the Sea was by far the most comprehensive set of negotiations in which the United States has participated. However, much of the management of individual resources or activities in international waters and in coastal waters is affected or accomplished by means of individual treaties or agreements. The United States is party to a number of international treaties or agreements which may directly or indirectly affect the management of its coastal waters.

For example, the United States has engaged in extensive bilateral negotiations with Canada on a variety of fishery-related issues. Among the subjects of negotiations have been international boundary lines for fishing and methods for managing inter-jurisdictional stocks. Agreements on such issues do, of course, affect to some degree the catch available to U.S. fishing interests, as well as the manner in which Federal Coastal Zone regulations are formulated and applied.

Furthermore, reacting to the slow progress of negotiations at UNCLOS III, the 96th Congress enacted "The Deep Seabed Hard Minerals Resources Act" (P.L. 96-283). Under this Act, the Administrator of the National Oceanic and Atmospheric Administration (NOAA), in consultation with the Secretary of State, is empowered to enter into negotiations with other states for the purpose of granting them reciprocal state status. In granting this status, the parties to the agreement consent to acknowledge the validity of licenses and permits granted by other parties to the agreement. Thus, by becoming party to a reciprocal state agreement, the United States would potentially limit the access of its own citizens to certain areas of ocean space while recognizing the rights of nationals of reciprocal states to exploit areas near its shores. By early 1984 the United States had concluded negotiations for a number of Reciprocating States Agreements. However, final signature has remained in abeyance largely because potential signatories have expressed reservations regarding the potential unfavorable reaction to such an agreement by the LOS Preparatory Commission.

## Technology Transfer in Multinational Negotiations

Furthermore, UNCLOS has not been isolated from other multilateral negotiations concerning technology transfer.[11] Implementation of the Law of the Sea Treaty will be influenced by negotiations that have been conducted in other UN-sponsored conferences on restrictive business practices, codes of conduct for the transfer of technology, and protection of patents and other industrial property. All of them supply information that may be used in the interpretation of the phrase "fair and reasonable commercial terms and conditions." In turn, the application of the technology transfer provisions of the Law of the Sea Treaty to the common property resources of the deep seabed can establish both a precedent for future negotiations in other resources beyond the limits of national jurisdiction and principles that can be used in the continuing discussions on the international transfer of technology.

## Technology Transfer and the "New International Economic Order"

The relevant points of the NIEO concerning technology transfer were enumerated during the Sixth Special Session of the UN General Assembly.[12] They included, inter alia:

(1) Raw materials and primary commodities in trade and development;
(2) International monetary system and financing of the development of developing states;
(3) Industrialization;
(4) Transfer of technology;
(5) Regulation and control over the activities of transnational corporations; and
(6) Charter of economic rights and duties of states.

The developing states considered technology important for two reasons. First, they produce much of the primary commodities and raw materials that are used by the developed states, but, without the proper technology, they are unable to share in the profits that result from downstream processing operations. Second, reliance on foreign sources of technology brings with it restrictions on the use of the technology which the developing states oppose for both economic and nationalistic reasons.

Transfer of technology is a particularly broad subject in the discussion between developed and developing states. Technology includes not only equipment but the training of personnel, patents and proprietary information, and the knowledge and ability necessary to modify and improve the equipment. While transfer of technology includes the sale of equipment, it may also encompass training programs, licenses to use patents, proprietary information and know-how, and other agreements between supplier and user. This was a major element of the negotiation in UNCLOS III.

## Development of Property Resources beyond the Limits of National Jurisdiction

Technology transfer became a major issue in UNCLOS because the developing states have recognized that access to technology is essential to their active participation in the exploitation of the mineral resources of the deep seabed. Furthermore, technology transfer issues are certain to arise as developing countries become directly interested in the exploitation of other resources beyond the limits of national jurisdiction including the mineral resources of Antarctica, the resources of the moon, the electromagnetic spectrum, and certain energy resources. The role of a Law of the Sea Treaty as a precedent for future resource management agreements was recognized by Ambassador Alvero de Soto, who served as coordinator for the Group of 77 at UNLCOS III. In September 1979, at a seminar on Antarctic resources, Ambassador de Soto said:

> The temptation to apply to Antarctica the principles which are the basis for the regime of the seabed is very great, and some have not been able to resist it. This temptation should increase now that the United Nations Committee on the Peaceful Uses of Outer Space has decided to recommend to the General Assembly a draft treaty that would declare that the Moon also constitutes the common heritage of mankind.

## NOTES

1. Sections of this chapter are adapted from Robert S. Jordan and Christopher Herrick, "Management of the Expanded Territorial Sea," a study prepared under a grant from the South Carolina Sea Grant Consortium and published by the Bureau of Governmental Research and Service, University of South Carolina.

2. Milner A. Ball, The Law of the Sea, Federal-State Relations and the Extension of the Territorial Sea (Athens, Georgia: Monograph #1, School of Law, University of Georgia, 1978), p. 7.

3. Ibid.

4. George A. Doumani, Ocean Wealth: Policy and Potential (Rochelle Park, New Jersey: Spartan Books, Hayden Book, 1972), p. 15.

5. Ball, Law of the Sea, p. 10.

6. As noted in Doumani, Ocean Wealth, p. 16.

7. As noted by Doumani (ibid., p. 18), the United States did grant oil leases to tracts of land over 100 miles off its shores under authority of this convention.

8. Various delegation reports, and the series of articles by Bernard H. Oxman et al. in the Journal of the American Society of International Law, provide relatively detailed descriptions of the work of the Conference by session.

9. As indicated in Ball (Law of the Sea, p. 19), 50 of the 119 coastal states have declared 12-mile territorial seas. An additional 30 states claim territorial seas of more than 12 miles width, while approximately 10 states claim territorial seas of between 3 and 12 miles breadth.

10. The difficulties experienced by the United States in supplying Israel during wartime seems to have been a significant factor in bringing this issue to the fore.

11. Adapted from unpublished paper by Lance Antrum and Christopher Herrick, "Issues of Technology Transfer in the United Nations Conference on the Law of the Sea," dated April 1980.

12. For a Third World perspective, see Robert S. Jordan, "Why a NIEO: The View from the Third World," in The Emerging International Economic Order: Dynamic Processes, Constraints and Opportunities ed. Harold Jacobson and Dusan Sidjanski (Beverly Hills: Sage, 1982).

# CONCLUSIONS: THE GLOBAL INTEREST OR THE NATIONAL INTEREST: MUST THEY LEAD TO CONFLICT OR CAN THEY LEAD TO COOPERATION?

*Robert S. Jordan*

## The Role of International Organizations in Shaping Policy

As already pointed out, there has been a marked increase over the past two decades in the number of newly independent Third World states and their participation at international conferences on economic and resources issues. These states have proposed such multilateral methods as resource cartels (OPEC), as commodities agreements (UNCTAD), loan deferrals (IMF), structuring resource regimes (LOS, IFAD, IEA), and transfers of technology (LOS, UN agencies, etc.) for dealing with North-South economic and developmental imbalances. These methods, of course, affect both developed and developing countries.

There have also been several attempts in the last ten years by global-oriented "trilateralists" and others in the United States, to maintain global economic stability under the domination of the industrial democracies of North America, Western Europe, and Japan. One of the mechanisms central to this task has been to include key Third World nations in those international conferences that were convened to negotiate the demands of the developing nations, or to formulate new acceptable international structures for an increasingly interdependent world. In contrast, thirty years ago, basic economic and resource decisions were made by a handful of industrially advanced states that were not compelled to be excessively concerned about the concerns and needs of the developing states of the Third World. At that time, there was no need for including developing countries in systemic-related negotiations. Even though the colonial ties still existed, the degree of economic interdependence was low unless interdependence in reality was dependency.

With the increase in the 1960s in both interdependency, and the number of newly independent, ideologically vocal, Third World states, international organizations such as IFAD, WFC, and UNCTAD and conferences such as UNCLOS became for a time in the 1970s major influences in determining resource access and expressions of Third World cohesiveness. Consequently, as we have seen, multilateral conferences have frequently been utilized to channel demands from the Third World for economic and resource transfer concessions from the "trilateral" advanced industrial states. The influential role of the Latin American, African, and Asian groups at UNCLOS III is testimony to this endeavor. Group cohesiveness at these conferences, as exhibited by the Group of 77, also frequently became a much more important source of influence than either individual states' national power or global status. This was especially true for the UN Special General Assembly Sessions and other NIEO-related ad hoc global conferences as well as the UNCLOS negotiations for an International Seabed Authority (even though in the final analysis, one state--the United States--was able to frustrate this outcome).

## General United States Foreign Policy Considerations in Multilateral Negotiations

Besides the implications for national security, the United States government's concern for protecting its perceived "vital" national interests in multilateral negotiations through ad hoc conferences or institutions, has obvious implications for general principles of United States foreign policy. These would include such principles as the promotion of friendly relations among states, the management of conflict through the development and codification of international conventional law, the strengthening of international institutions so as to provide machinery for settling disputes over as well as for financing and allocating resources, and the provision of some rational consistency in cost-benefits between short-term domestic economic concerns and the long-term international economic welfare. The United States has been committed in varying degrees to upholding the spirit and letter of these principles, and has reaffirmed this commitment by ratifying various international instruments, including the United Nations Charter.

In this respect, the United States supported the work of the United Nations' International Law Commission in the 1950s in preparing draft articles on the Law of the Sea as well as UNCLOS I and UNCLOS II, held in Geneva in 1958 and 1960. Subsequently, the United States also supported the 1970 General Assembly Resolution 2749 on the Declaration of Principles on the Peaceful Uses of the Seas Beyond the Limits of National Jurisdiction, and the convening of UNCLOS III in 1973. Four Presidents and their Administrations supported the conclusion of a comprehensive Treaty on the Law of the Sea because they judged that such a Treaty would serve American national security as well as foreign policy interests. The Reagan Administration's policy review and its objections to the seabed mining provisions of the Draft Convention thus raised in the minds of many governments serious questions about the commitment of the United States to friendly relations among states and the rule of the law in general, and to the rules of Conference procedure in particular. In the latter respect, at the first and second UNCLOS sessions, the United States and other Conference participants agreed to a novel decision-making arrangement, a "gentlemen's agreement," that consisted of three elements:

1. The principle of consensus whereby a proposed treaty provision would be accepted only if it enjoyed widespread consensus of Conference participants: similarly, a treaty provision once accepted by consensus could only be altered if the alternative proposed provision enjoyed wide consensus.

2. A procedure for voting, only if and when the possibility of obtaining consensus had been entirely exhausted.

3. The concept of a comprehensive treaty (or package deal) consisting of diverse, interlocking parts which the Conference would examine upon conclusion of negotiations.

This decision-making procedure gained wide legitimacy in the Conference because officers and government representatives continued to endorse it in subsequent sessions of the Conference.

When the United States set forth its objections to the seabeds text, several delegations recalled that the United States had committed itself to the "gentlemen's agreement" and that the compromises that had been reached in these negotiations should be adhered to and not reopened "to secure further advantages for a particular country in pursuit of its national interest." Nonetheless, as we have seen, the Reagan Administration pursued another, more unilateral, course of action.

## Implications for the United States of United Nations "Conferencing" over Resource Issues

Because there has been an increase in the number of Third World states participating in international conferences, we have already observed that the bargaining position of the United States has been weakened in international forums in recent decades. The diminishing American power to control unilaterally conference outcomes can be seen in the many concessions the United States made on deep seabed mining at UNCLOS III to the Group of 77, and the United States' reluctance at the World Food Conference to rely on FAO to resolve food issues, instead promoting the creation of new, American-dominated international food institutions--the World Food Council (WFC) and the International Fund for Agricultural Development (IFAD). When the global NIEO conferences began in the 1970s, the full importance of resource accessibility as well as ownership was just beginning to be fully appreciated, OPEC's bombshell to the contrary notwithstanding. As that appreciation increased during the course of the decade, as regards energy, marine minerals, and food, possible multilateral means of allocating resource accessibility or ownership appeared more and more a mistake in purely economic terms, and hence American confidence in such Third World-oriented United Nations agencies as UNCLOS, IEA, and FAO continued to diminish (UNESCO being the most recent casualty).

Of major importance in coming to terms with the highly interdependent nature of international resource politics, then, is the need to understand more fully the role played by international organizations in multilateral resource management. The international organization model offers an analytical framework for understanding the complexities involved in multilateral resource management by focusing on the role and significance of international state and nonstate actors and their cross-national interdependencies. If international organizations are not used by the United States at least for negotiating differences, if not for carrying out resource financing or allocation, then there may well be increasing instances of individual states settling their resource or economic differences through the use of force.

Unfortunately, the likelihood of the advanced industrial states successfully working together to ensure access to resources is not very high. For example, during the Middle East oil embargo of 1973-74, and the years which followed, the OECD member-states were unable to plan and to operationalize a united stance against OPEC political pressure because of the cartel's monopoly of oil supply to them. (The major oil companies, however, did effectively work out distributional problems of the 1973-74 oil boycott by reallocating oil while at sea.)

In summary, on the one hand, the advanced industrial states have been attempting to negotiate resource policy in different international forums--both official and unofficial. On the other hand, individual states, including the United States, are finding such negotiations and international organizations inadequate in providing political and other guarantees. These states are, therefore, often pursuing bilateral (or "one-on-one") arrangements. The Reagan Administration in particular has turned in this direction.

For example, the World Bank has based its view of how to deal with Third World food shortages on need. It has made Africa its regional priority for the 1980s and has increased its disbursements on the continent by 50 percent (to over $1.1 billion plus a $2 billion increase in economic assistance). In contrast, the Reagan Administration justifies its aid on ideology and East-West considerations--although aid increased 40 percent for 1981-84, arms sales and assistance increased 150 percent. Furthermore, the Administration has emphasized bilateral aid based on free-market and growth-oriented policies that are difficult to justify either on humanitarian grounds or on the true nature of the economies of most of the states of Africa.

Multilaterally, the Administration has opposed fifty loans to Africa considered by the World Bank or smaller multinational development agencies--among them six loans for starving Ethiopia that were proposed in 1984. The Administration has also refused to increase fiscal 1985 contributions to the World Bank's International Development Association (the soft-loan affiliate). It is also cutting back on donations to IFAD.[2]

## Trends That Could Affect the United States in Multilateral Resource Management

Thus, certain trends have emerged which could have adverse implications for the United States. The first has been an erosion of "global-interest" and "international-interest" themes in favor of more specific "special-interest" and "national-interest" proposals. This is well illustrated by the ever-expanding territorial claims of coastal states. Because this trend has been legitimized in the Law of the Sea Treaty, the original theme of the "common heritage of mankind," which was based on the concept of narrow territorial waters and narrow national economic zones, has been all but lost to the national and special-interest scramble for greater access to the resources of the ocean.

A second trend concerns the continuing conflict between advanced industrial and developing states. The degree of the developing states' distrust of United States foreign policy initiatives has become very strong within those international organizations which were designed for negotiating international issues. As a result, initiatives of the United States are frequently rejected automatically by coalitions of the developing states acting as a remarkably solid unit on many issues, or, where the United States "punishes" a state for not voting its way in an international forum by withholding aid or other benefits, deep-seated resentments are built up that reinforce confrontation rather than conciliation. Furthermore, many developing states with coastal waters abundant with fisheries and continental margins rich in minerals have successfully exploited the overall Third World's distrust of United States policy initiatives to compromise themselves the common heritage principle.

But UNCLOS III and OPEC have also demonstrated a third trend: the tentative nature of Third World cohesion. Many times cohesiveness of the Group of 77 has been stressed to the breaking-point over several issues. This was particularly so over the questions of the right of land-locked and shelf-locked Third World states to some of the benefits of other developing states' coastal economic zones. It also has been true in attempts within OPEC both to agree upon, and then to comply with, production quotas and prices.

A fourth trend, implicit in the first three, concerns the extraordinary difficulties of reaching optimal solutions, or any solution for that matter, to the problems of multilateral resource management. UNCLOS III made clear the difficulties inherent in a conference where over 150 states attempted to negotiate an international issue with hundreds of distinct sub-issues. It has also demonstrated several problems with the present mode of international negotiations. For instance, domestic interest groups frequently formed both national and transnational alliances to support, modify, or defeat national positions on specific issues. As in all foreign policy issues, it is therefore necessary to reconsider the linkages between domestic interest groups, national governments, and the process of multilateral diplomacy in general. The sliding influence[3] of OPEC is also worth examining in this respect.

Yet, despite the limitations of international organizations, they do offer hope that conflict can be reduced by offering the opportunity to discuss issues of increasing international interdependence in a global forum. In this respect, governments of all nations will be looking closely at the ultimate success or failure of the Law of the Sea Treaty process, the UNCTAD process, and the OPEC process, among others.

## Changes in the Multilateral Management of Food

There are two ways in which the understanding of the problems of hunger and malnutrition prevalent at the time of the 1974 World Food Conference changed significantly in the intervening decade:

(1) Hunger and malnutrition are now viewed as global realities, no longer simply as Third World problems.

(2) Hunger and malnutrition have come to be understood in less abstract and more locale-specific terms.

As food trade increased among states, agricultural self-reliance was dealt a serious blow in many states. Foodstocks in the United States in particular, but not exclusively, have become global supermarkets. As a result, as discussed earlier, many states have increased their dependence on the international market place for export earnings in order to pay for their food imports.

But very little progress has been made in assuring more stable agricultural prices, in increasing the export earnings of food-importing Third World states, or in otherwise achieving more balanced world trade patterns. It would therefore make good sense to moderate any internationally disruptive domestic agricultural policies of the food-exporting states and to use current agricultural abundance to enhance the food security of food-importing Third World states. To do this would require the use of international machinery--multilateral means which are already available but not fully utilized.

The commonality of interests between producers in advanced industrial and developing states should be more fully understood and should provide a firmer basis for global and national agricultural policies. For example, some of the agricultural policies of the advanced industrial states that were designed with export promotion as a centerpiece (and this is increasingly the case for the United States) pit their producers, who must rely on exports for their incomes, against Third World producers who cannot compete with these highly subsized imports. It is crucial, therefore, for both kinds of producers, that these policies be adjusted.

The World Food Conference was correct--even clairvoyant--in endorsing the view that the increasing interdependence of the economies of individual states necessitates a "global concept of agricultural adjustment" which promotes prices that are "remunerative to producers and fair to consumers" (Resolution XIX). While in the current situation some farmers and farm organizations in some advanced industrial states stress agricultural export promotion at the expense of Third World agricultural producers and production, what has been termed the "inward looking stance" of advanced industrial states' agricultural policies can, taking a longer view, be modified to permit some of the "painful adjustments" needed in their domestic farm sectors. The policies of the United States in this respect are crucial although, as far as the Reagan Administration is concerned, not promising.

Experience indicates that as the Third World grows more of its own food, the accompanying overall economic growth makes for better markets for food and other imports. Highly subsidized agricultural exports which, in effect, undermine Third World agriculture will not in the long term serve the interest of producers in either the North or the South. American and other advanced industrial state producers, themselves concerned about widespread hunger in many areas of the world, would rather produce than allow their fields to remain idle. At the same time, they also should support a more equitable division of production with Third World producers--provided that such adjustments are made within a policy framework that provides a fair income for both sets of farmers.

The challenges posed both to national (bilateral) and multilateral institutions in coping with mass starvation and crippling malnutrition, accompanied by desertification, in Africa, can provide an opportunity for reinforcing or creating more effective multilateral means for food resource management.

There is also a constructive role that nongovernmental organizations (NGOs) can play in stressing a multilateral perspective as well as a human focus in trade, agricultural adjustment, and other similar hunger issues. As people-to-people organizations with ties across national boundaries, NGOs can bridge existing gaps and can promote solidarity among producers in states whose governments may not necessarily be attuned to their common interests. NGOs can also play a role in educating farmers and other constituencies on global aspects of the hunger issue. In fact, it will be recalled that NGOs were active at the World Food Conference, as they have been since, in encouraging the framing of hunger issues in global rather than food-export versus food-import terms.[4]

Finally, NGOs can also make a contribution of a more operational sort in trade matters. They can help in identifying external markets for Third World exports and in stimulating within states more responsible market mechanisms as, for example, through the establishment and strengthening of cooperatives. In all of this, NGOs are committed to help assure fair returns to Third World producers from international and external trade.

## Factors for Success of Other Resource Cartels as a Form of Multilateral Management

The success of OPEC in 1973 led to a revolution in the thinking of other Third World raw material producers. Suddenly, it seemed that producer cartels could bring the end of dependence. Producer organizations in copper, bauxite, iron ore, bananas, and coffee were either formed or took on a new life. As we have seen in April 1974 at the Special Session of the United Nations General Assembly, the Group of 77 called for the establishment of a New International Economic Order which was to include support from the advanced industrial states for the functioning and aims of producer associations. In December 1974 the Third World pressed through the United Nations General Assembly the Charter of Economic Rights and Duties of States, which included an article on the "right" of the Third World states to form producer associations and the "duty" of the advanced industrial states to "respect that right by refraining from applying economic and political measures that would limit it."

Yet by the 1980s the prospects for these new producer cartels seemed dim indeed. Virtually none, including OPEC, had succeeded in maintaining higher commodity prices in the face of depressed market conditions; most were fraught with internal dissension; and a few never really got off the ground. Still, the idea of raw material producer associations is far from dead, and it is worthwhile to examine both their potential and the reasons for their very limited successes to date.

In the aftermath of OPEC's initial success, several factors seemed to suggest that in the short term, and perhaps in the medium term, some producer cartels might have succeeded. In the near term, economic conditions were propitious for many commodities, particularly for those on which the advanced industrial states were highly dependent. The United States, for example, relies on imports of bauxite, tin, bananas, and coffee. Western Europe and Japan, less well-endowed with raw materials, depend also on imports of copper, iron ore, and phosphates. The disruption of supply of many of these commodities, particularly critical minerals, could have an important disruptive effect on their developed market economies.

Tight market conditions favor producers over the short term. As demonstrated by the 1973-74 oil action, and possibly in regard to food, as well as minerals, a seller's market facilitates cartel action by enabling one or a small number of producers to raise prices. An economic boom coupled with uncertain currency markets can encourage speculation in commodities that in turn can lead to commodity shortages with consequent price increases. For example, during this period, Morocco in phosphates and Jamaica in bauxite took advantage of this situation to raise prices.

In addition to these economic factors, several political conditions can also favor producer action, again primarily over the short run. For many commodities--for example, bauxite, copper, phosphates, bananas, cocoa, coffee, natural rubber, and tea--a relatively small number of Third World producers dominate the export market, and a number of these producers have formed associations with the goal of price management.

## Political Factors Favoring Multilateral Resource Management through Cartels

Several political developments made multilateral producer cooperation more likely in the mid-1970s. One was a new sense of Third World self-confidence. The OPEC experience suggested to other Third World producers that, through their control of commodities vital to the advanced industrial states, they might possess the threat they had long sought. Thus many Third World states felt that they could risk more aggressive policies. Another new development stemmed not from confidence but from desperation. As we know, the simultaneous energy, food, recession, and inflation crises left most Third World states with severe balance-of-payment problems. Some of these states may have felt that they had no alternative to instituting risky measures which might provide short-term economic benefits, but which would probably prove unsuccessful or even damaging in the long run.

Reinforcing this sense of economic desperation was political concern. Political leaders, especially those in the Third World, tend to have short-run perspectives, as maintaining themselves in power may depend on achieving short-term gains despite inevitable long-run losses. However, as discussed earlier, this argument is directly opposite to the OPEC model for a successful producer cartel, wherein for a time monetary reserves enabled the producing states to take a long-term perspective, to risk short-term losses for long-term gains. In other cases, producers with huge balance-of-payments deficits may be moved to risk long-term losses for short-term gains. And, as has been argued, short-term maximization of revenues may in fact be rational action for the long-term view; that is, if producers feel that their short-term profits will be sufficient to achieve economic diversification and development, they may rationally pursue short-term gains that would reap political advantages.

Finally, political cooperation was sometimes facilitated by the nature of the task of managing price and supply. In such commodities as bauxite and bananas, vertically integrated oligopolistic multinational corporations could be taxed according to the OPEC formula. In these and other commodities, production control was facilitated by increasing governmental regulation or ownership of production facilities.

With all of these factors working in favor of cartel success, why then have the raw material producer associations been so <u>unsuccessful</u> since 1974? Some of the reasons for the cartels' problems can be traced to the depressed economic conditions of the late 1970s and early 1980s; others are of a more general nature.

The successful survival of a cartel generally depends on two complex factors. First, producers have to manage price so that it does not rise above a level which would encourage use of substitutes. Such management requires sophisticated market knowledge and predictive ability. Then, too, because the threshold price may be lower than the preferred price for many producers, agreement on joint action may be quite difficult to achieve. Second, and equally difficult, the supply response from other producers must be managed. Currently existing cartels have been generally unable to manage successfully either prices or supply; price-cutting among fellow cartel members has been common, and few emerging producers have agreed to supply controls.

For, despite some incentive for cooperation, there have been major problems of multilateral management. While many commodities are supplied by a small number of producers, these producers often find they have more in conflict than in common. The copper producers, for example, are divided by political as well as economic differences. Moreover, while a foreign exchange crisis may encourage cooperation, it also may facilitate consumer resistance. Third World producing states that have no reserves and that rely on the export of one commodity for the bulk of their foreign exchange earnings are not in a position to endure long concerted corporate or consumer government resistance. Furthermore, the temptation to take short-term profits from concerted action at the expense of longer-term gains is greater in situations of balance-of-payments crisis. And, while the task of price management may be easy in some cases, as when there is a leader and when multinational corporations are present, such advantages do not exist for many commodity-producing nations.

One of the greatest barriers to producer cartels has been the task of managing supply. Few states have a large enough share of production and large enough reserves to engage in the kind of leadership role played by Saudi Arabia. No one state or small group of states is able to bear the burden of supply reduction for the entire commodity group.

## Buffer Stocks, Managed Multilaterally, as a Possible Solution

In the absence of tight markets, then, supply can only be controlled through buffer stock schemes or through export or production reductions--methods that are politically complex and economically costly. Many food and other commodities are perishable and hence cannot be stored in a buffer stock, whereas other commodities require enormous buffer stocks and vast financing to maintain prices. Export and production reductions are equally difficult to accomplish. Export reduction without production controls poses the same problems of storing and financing as buffer stocks. And multilateral agreements to reduce production are difficult to achieve--even OPEC has never been able to agree on a prorationing scheme--and may be costly in terms of domestic employment.

Perhaps the most devastating blow to the producer associations has been dealt by the stagnant--in some cases declining--demand for their commodities. In 1974 and 1975, as economic activity in the industrialized states declined, demand for industrial raw materials fell precipitously. Since then economic uncertainty in the advanced industrial states has meant instability in raw material markets. For example, world consumption of refined copper dropped from 8.8 million metric tons in 1973 to 7.5 million metric tons in 1975; American consumption dropped by more than one-third. Since then world demand for copper has remained stagnant, never surpassing 1973 levels.[5]

Faced with reductions or slow growth in the demand for their commodities, the only hope for the Third World producer associations would have been to reduce production and supply in order to maintain prices at desired levels. Yet, as we have indicated, most producing states have found it politically or economically difficult to cut back production, and many cartel members have engaged in price-cutting to increase their international competitiveness (i.e., OPEC). The result has been a general oversupply of many raw materials and a downward pressure on raw-material prices which the cartels have been unable to counteract.

## The Role of Government Agencies in Multilateral Management

Another aspect of multilateral management is that some governmental agencies' interests are, at times, more like those within another state than like those of their own competing domestic agencies. For example, the agencies involved with securing access to the deep seabed in the United States, France, Great Britain, Japan, and West Germany frequently have had more in common than they have with ministries within their respective governments seeking to ensure, through the LOS Treaty, free passage through international straits. UNCLOS III and other international conferences provide the physical contact and legitimacy for some of these potential transgovernmental coalitions to become active.[6] The fallback positions of the United States in bargaining were often disclosed in advance at UNCLOS III negotiations, and the disarray in the United States delegation in the World Food Conference revealed only too clearly the policy disarray in Washington. As discussed earlier, there had been, for instance, lobbying by the Interior Department and oil company officials at UNCLOS with those Third World states favoring broad coastal state jurisdiction over the continental shelf (contrary to then official United States policy). In addition to lobbying governments at international conferences, transnational oil and mining companies have frequently formed joint ventures with partners from several states to influence their governments' perceptions and definitions of their interests in new resource policy areas. An obvious example of this, outlined earlier, was the success of American and British oil companies in the Middle East up to 1973 in influencing the oil policies of the OPEC nations. Another example is the success of American mining companies in several Latin American countries (such as Chile) up to the early 1970s in influencing resource policy decisions and their internal political ramifications.

## From Crisis Diplomacy to Crisis Management

From the foregoing, it is quite apparent that there is no clear line that separates a "national interest" from a "global interest." In food, energy, and marine minerals, even if interdependence has not taken hold as a concept of public policy at the national level, in none of these issue areas can any state--including the United States--work out its policies in isolation from those other states that are affected by them. Sooner or later unilateralism or bilateralism must give way to some form of multilateralism. This may not take the form of "management" as commonly understood in large public or private bureaucracies, but nonetheless the "management" of these issues as the term is used in "crisis management" is not inappropriate.

Harlan Cleveland prescribed certain steps in what he termed "crisis diplomacy" that can apply to the management of such potential crises as conflict over food, energy, or marine minerals. But first he had observed:

> . . . the problem of decision-making in our complicated world is not how to get the problem simple enough so that we can all understand it; the problem is to get our thinking about the problem as complex as humanly possible--and thus approach (we can never match) the complexity of the real world around us.[7]

He listed as "Lesson No. 1: Keep Your Objectives Limited." Certainly in the debate between the advanced industrial states and the Third World over a New International Economic Order, the South did not observe this rule, and suffered the consequences: stalemate. This leads to "Lesson No. 2: Decide How Far You Would Go." The early success of OPEC provided what subsequently turned out to be a poor example for the Third World to apply in forming other mineral cartels--transference was not possible. The "realities" of the global marketplace resisted rendering something extremely complex to the relative simplification of a cartel. Furthermore, unfortunately for the hungry and undernourished of the world, the food-consuming states and the producing states seem unable to agree on how far to go to meet their respective needs without rearranging the dangerous dependency relationship that has grown up in the past twenty years. Some form of multilateral crisis management is required that seems incapable of being brought into being, the consequence being a tremendous, unfortunate potential for widespread suffering that could lead to more and more situations of relatively low-level but persistent conflict. This leads us to "Lesson No. 3: Creep Up Carefully on the Use of Force." Certainly in energy, the use of force has been in the forefront of the strained relationship between the producers and the consumers. However, the "controlled political use of force" is essential if widespread tensions do not lead to catastrophic consequences. The failure to utilize effectively multilateral means to isolate energy as a tool of short-term national interest may come back to haunt the world just as the failure of the diplomats of 1914 to avoid using military mobilizations as a tool of short-term national interest led to catastrophe for all. Finally, "Lesson No. 4: Widen the Community of the Concerned" applies directly to the marine minerals issue. Even if the formulations of the Law of the Sea Treaty proved in the last analysis unacceptable to the United States, some method other than boycott or forming a counter-bloc (which has so far fortunately failed) must be employed that will permit further opportunities for multilateral negotiations that could lead to the multilateral management for all states of these potentially vital resources.[8]

## Notes

1. In preparing this chapter, the following sources have been used, acknowledgment of which is made here:

   Harlan Cleveland, "Crisis Diplomacy," Foreign Affairs (July 1963): 3-14.

   Margaret E. Galey, "Foreign Policy Implications of Seabed Mining," unpublished manuscript, September 17, 1981.

   John P. Renninger, "Global Negotiations and the United Nations," an unpublished paper prepared for the 24th Annual Convention of the International Studies Association, Mexico City (April 5-9, 1983).

   Kurt Michael Shusterich, Resource Management and the Oceans: The Political Economy of Deep Seabed Mining (Boulder, Colorado: Westview Press, 1982), pp. 138-141.

   Joan Edelman Spero, The Politics of International Economic Relations, 2d ed. (New York: St. Martin's Press, 1981), pp. 270-276.

   The World Food and Hunger Problem: Changing Perspectives and Possibilities, 1974-1984, an independent assessment presented to the World Food Council (UNWFC Doc. WFC/1984/6, 10 February 1984.)

"Toward Redoubled Efforts to End Hunger and Malnutrition: Some Non-Governmental Organization Recommendations," a report prepared for the Tenth Ministerial Session of the World Food Council, 11-15 June 1984.

Thomas G. Weiss and Robert S. Jordan, The World Food Conference and Global Problem Solving (New York: Praeger, 1976).

2. Drawn from Jack Shepher, "The Next Big Step to Help Africa," New York Times, 10 January 1985.

3. See, for example, Youssef M. Ibrahim, "OPEC's Old Firm Grip on World's Oil Prices Becomes Ever Weaker," The Wall Street Journal, 11 January 1985.

4. See, for example, Robert S. Jordan, Interest-Group Politics in Global Conferencing: An Assessment of the Contemporary Role of Nongovernmental Organizations in Intergovernmental Organizations, Second Series No. 5, Occasional Papers, Section on International and Comparative Administration, American Society for Public Administration, 1984.

5. A recent report of the Office of Technology Assessment recommended that the United States should reduce its dependence on the Soviet Union and South Africa for chromium, cobalt, manganese, and platinum--materials vital to defense needs (reported in USA Today, 15 January 1985). By increasing domestic production, improving conservation and/or finding substitutes, the United States can reinforce its unilateral interests, but this does not, of course, encourage multilateralism as an alternative to a bilateral demand-supply relationship.

6. See, for example, Weiss and Jordan, World Food Conference, esp. under Ch. 4, "The World Food Conference as Model for Global Discussion."

7. Cleveland, "Crisis Diplomacy."

8. Ibid., p. 3.

# SELECTED BIBLIOGRAPHY

## ENERGY

### Books

Adelman, M. A., et al. Oil, Divestiture, and National Security. New York: Crane, Russak, 1977.

Al-Chalabi, Fadhil J. OPEC and the International Oil Industry: A Changing Structure. New York: Oxford University Press, 1980.

Alnasrawi, Abbas. Arab Oil and United States Energy Requirements. Belmont, MA: Association of Arab-American University Graduates, 1982.

Anderson, Irvine H. Aramco, the United States and Saudi Arabia: A Study of the Dynamics of Foreign Oil Policy. Princeton, NJ: Princeton University Press, 1981.

Aperjis, Dimitri. The Oil Market in the 1980s: OPEC Oil Policy and Economic Development. Cambridge, MA: Ballinger, 1982.

Belgrave, Robert. Oil Supply and Price: Future Crisis Management. The Atlantic Papers, no. 48. Paris: Atlantic Institute for International Affairs, 1982.

Blair, John M. The Control of Oil. New York: Pantheon Books, 1976.

Bohi, Douglas R. Limiting Oil Imports: An Economic History and Analysis. Baltimore: Published for Resources for the Future by the Johns Hopkins University Press, 1978.

Bohi, Douglas R., and Montgomery, W. David. Oil Prices, Energy Security, and Import Policy. Washington, D.C.: Resources for the Future, 1982.

Brossard, E. B. Petroleum--Politics and Power. Tulsa, OK: Pennwell Books, 1983.

Bundy, William P., ed. The World Economic Crisis. New York: Norton, 1975.

Chester, Edward W. United States Oil Policy and Diplomacy: A Twentieth Century Overview. Westport, CT: Greenwood Press, 1983.

Choucri, Nazli, and Ferraro, Vincent. International Politics of Energy Interdependence. Lexington, MA: Lexington Books, 1976.

Conant, Melvin. The Oil Factor in the U.S. Foreign Policy, 1980-1990. Lexington, MA: Lexington Books, 1982.

Congressional Quarterly, Inc. Continuing Energy Crisis in America. Washington, D.C.: Congressional Quarterly, 1975.

Crist, Meredith S., and Laffer, Arthur B., eds. Future American Energy Policy. Lexington, MA: Lexington Books, 1982.

Engler, Robert. The Brotherhood of Oil: Energy Policy and the Public Interest. Chicago: University of Chicago Press, 1977.

Franssen, Herman, et al. World Energy Supply and International Security. Cambridge, MA: Institute for Foreign Policy Analysis, 1983.

Ghosh, Arabinda. OPEC, the Petroleum Industry, and United States Energy Policy. Westport, CT: Quorum Books, 1983.

Goldstein, Donald J., ed. Energy and National Security: Proceedings of a Special Conference. Washington, D.C.: National Defense University Press, 1981.

Horwich, George, and Mitchell, Edward J., eds. Policies for Coping with Oil-Supply Disruptions. Washington, D.C.: American Enterprise Institute for Public Policy Research, 1982.

Hurewitz, J.C., ed. Oil, the Arab-Israel Dispute, and the Industrial World. Boulder, CO: Westview Press, 1976.

Jaidah, Ali M. An Appraisal of OPEC Oil Policies. New York: Longman, 1983.

Jones, Aubrey. Oil, the Missed Opportunity: Or Naft and Shaft. London: Deutsch, 1981.

Kash, Don E., and Rycroft, Robert W. U.S. Energy Policy: Crisis and Complacency. Norman: University of Oklahoma Press, 1984.

Katy, James Everett. Congress and National Energy Policy. New Brunswick, NJ: Transaction Books, 1984.

Kaufman, Burton I. The Oil Cartel Case: A Documentary Study of Antitrust Activity in the Cold War Era. Westport, CT: Greenwood Press, 1978.

Klebanoff, Shoshana. Middle East Oil and U.S. Foreign Policy, with Special Reference to the U.S. Energy Crisis. New York: Praeger, 1974.

Knowles, Ruth S. America's Energy Famine: Its Cause and Cure. Norman: University of Oklahoma Press, 1980.

________. America's Oil Famine: How It Happened and When It Will End. New York: Coward, McCann & Geoghegan, 1975.

Kokxhoorn, Nicoline. Oil and Politics: The Domestic Roots of U.S. Expansion in the Middle East. Frankfurt am Main: P. Lang, 1977.

Krueger, Robert B. The United States and International Oil: A Report for the Federal Energy Administration on U.S. Firms and Government Policy. New York: Praeger, 1975.

Landis, Robin C., and Klass, Michael W., eds. OPEC: Policy Implications for the United States. New York: Praeger, 1980.

Landsburg, Hans H., et al. Energy--The Next Twenty Years. Cambridge, MA: Ballinger, 1979.

Lawrence, Robert, ed. New Dimensions to Energy Policy. Lexington, MA: Lexington Books, 1979.

Lawrence, Robert M., and Heisler, Martin O. International Energy Policy. Lexington, MA: Lexington Books, 1980.

Levy, Walter J. Oil Strategy and Politics, 1941-1981. Boulder, CO: Westview Press, 1982.

Lieber, Robert J. The Oil Decade: Conflict and Cooperation in the West. New York: Praeger, 1983.

Lonnroth, Mans, Steen, Peter and Johansson, Thomas. Energy in Transition: A Report on Energy Policy and Future Options. Berkeley: University of California Press, 1980.

Lund, Leonard, ed. Energy: Update and Outlook, November 1974. New York: Conference Board, 1975.

Mancke, Richard B. Squeaking By; U.S. Energy Policy since the Embargo. New York: Columbia University Press, 1976.

McGovern, James. The Oil Fame. New York: Viking Press, 1981.

Medvin, Norman. The Energy Cartel: Who Runs the American Oil Industry? New York: Vintage Books, 1974.

Medvin, Norman, et al. The Energy Cartel: Big Oil vs. the Public Interest. New York: Marine Engineers' Beneficial Association, 1975.

Mendershausen, Horst, and Nehring, Richard. Protecting the U.S. Petroleum Market against Future Denials of Imports. R-1603-ARPA. Santa Monica, CA: Rand, 1974.

Miller, Robert L. The Economics of Energy: What Went Wrong? New York: Morrow, 1974.

Newlon, Daniel H., and Breckman, Norman V. The Oil Security System: An Import Strategy for Achieving Oil Security and Reducing Oil Prices. Lexington, MA: Lexington Books, 1975.

Noreng, Oystein. Oil Politics in the 1980s: Patterns of International Cooperation. New York: McGraw-Hill, 1978.

Odell, Peter R., and Vallenilla, Luis. The Pressures of Oil: A Strategy for Economic Revival. New York: Harper & Row, 1978.

Oil Diplomacy: The Atlantic Nations in the Oil Crisis of 1978-79. Philadelphia: Foreign Policy Research Institute, 1980.

Pfaltzgraff, Robert L. Energy Issues and Alliance Relationships: The United States, Western Europe and Japan. Cambridge, MA: Institute for Foreign Policy Analysis, 1980.

Plummer, James L., ed. Energy Vulnerability. Cambridge, MA: Ballinger, 1982.

Rabe, Stephen G. The Road to OPEC: United States Relations with Venezuela, 1919-1976. Austin: University of Texas Press, 1982.

Rand, Christopher T. Making Democracy Safe for Oil: Oilmen and the Islamic East. Boston: Little, Brown, 1975.

Ridgeway, James, and Conner, Bettina. New Energy: Understanding the Crisis and a Guide to an Alternative Energy System. Boston: Beacon Press, 1975.

Rustow, Dankwart A. Oil and Turmoil: America Faces the Middle East. New York: Norton, 1982.

Safer, Arnold E. International Oil Policy. Lexington, MA: Lexington Books, 1979.

Sampson, Anthony. The Seven Sisters: The Great Oil Companies and the World They Made. New York: Viking Press, 1975.

Sawyer, Herbert L. Soviet Perceptions of the Oil Factor in U.S. Foreign Policy: The Middle East-Gulf Region. Boulder, CO: Westview Press, 1983.

Shaffer, Ed. The United States and the Control of World Oil. New York: St. Martin's Press, 1983.

Sherrill, Robert. The Oil Follies of 1970-1980: How the Petroleum Industry Stole the Show (and Much More Besides). Garden City, NY: Anchor Press/Doubleday, 1983.

Sobel, Lester A., comp. Energy Crisis. Facts on File, 1974, 3v.

Solberg, Carl. Oil Power. New York: Mason/Charter, 1976.

Specter, Herschel. Getting the West Out of the Oil Dilemma: An Energy Family Approach. Washington, D.C.: Atlantic Council of the United States, 1979.

Stobaugh, Robert, and Yergin, Daniel, et al., eds. Energy Future: Report of the Energy Project of the Harvard Business School. New York: Random House, 1979.

Szulc, Tad. The Energy Crisis. New York: Watts, 1974.

Szylowicz, Joseph S., and O'Neill, Bard E., eds. The Energy Crisis and U.S. Foreign Policy. New York: Praeger, 1975.

Szylowicz, Joseph S., and O'Neill, Bard E., eds. Petropolitics and the Atlantic Alliance. Washington, D.C.: National Defense University. Research Directorate, 1976.

Wyant, Frank R. The United States, OPEC, and Multinational Oil. Lexington, MA: Lexington Books, 1977.

Zinberg, Dorothy S., ed. Uncertain Power: Struggle for a National Energy Policy. New York: Pergamon Press, 1983.

## Journals

Abu ad Khail, Mohammed. "The Oil Prince in Perspective." International Affairs (London) (October 1979):517-30.

Amuzegar, Jahangir. "A World Without OPEC." Washington Quarterly 5 (Autumn 1982):62-70.

Bagnaso, Alberto M. "Oil and Money: A Note on the Western Financial Community." Orbis 23 (Winter 1980):875-96.

Beenstock, Michael. "OPEC and the World Economy." Year Book of World Affairs 36 (1982):96-110.

Brown, William R. "The Oil Weapon." Middle East Journal 36 (Summer 1982):301-18.

Cherneny, H.B. "Restructuring the World Economy: Round II." Foreign Affairs (Summer 1981): 1102-20.

Choucri, Nazli. "Analytical Specifications of the World Oil Market: A Review and Comparison of Twelve Models." Journal of Conflict Resolution 23 (June 1979):346-72.

Conradi, Alberto Quiros. "Energy and the Exercise of Power." Foreign Affairs (Summer 1979): 1144-66.

Cooper, Mary H. "OPEC: 10 Years after the Arab Oil Boycott." Editorial Research Reports (23 September 1983):699-720.

Dafter, Ray. "World Oil Production and Security of Supplies." International Security (Winter 1979-80):154-76.

Doran, Charles F. "OPEC Structure and Cohesion: Exploring the Determinants of Cartel Policy." Journal of Politics 42 (February 1980):82-101.

Emerson, Steven. "The American House of Saud: II, the Aramco Connection." The New Republic (19 May 1982):11-16.

Erb, Richard Q., "The Gulf Oil Producers: Overview and Oil Policy Implications." AEI Foreign Policy and Defense Policy Reviews 2, nos. 3-4 (1980).

________. "The Organization of Petroleum Exporting Countries Members and U.S. Policies." Annals of the American Academy of Political and Social Science 460 (March 1982):111-17.

Gholamnezhad, Abdolhamid. "Critical Choices for OPEC Members and the United States." Journal of Conflict Resolution (March 1981):115-43.

Giritli, Ismet. "Thoughts on OPEC." Annals de la Faculte de Droit d'Istanbul 42 (1979):545-54.

Goldmuntz, Lawrence A. "Toward a New U.S. Oil Policy." Washington Quarterly (Spring 1979): 54-59.

Hall, Robert E., and Pindyck, Robert S. "Oil Shocks and Western Equilibrium: Rising Oil Prices Don't Have to Threaten the Economic Prosperity of the Industrial Nations if They Adopt the Right Policies." Technology Review (May-June 1980):32-40.

Hansen, Herbert E. "OPEC's Role in a Global Energy and Development Conference." Journal of Energy and Development (Spring 1980):182-93.

Hansen, John R., and Reynolds, Morgan O. "Encouraging Future OPECs." Policy Review 12 (Spring 1980):123-32.

Healey, Denis. "Oil, Money and Recession." Foreign Affairs 58 (Winter 1979-80):217-30.

Jawdat, Nameer Ali. "The Recurring Energy Crisis." Washington Quarterly 5 (1982):71-75.

Jordan, Amos A. "Energy and National Security: Sizing Up the Risks." Washington Quarterly (Summer 1980):154-63.

Katz, James E. "The International Energy Agency: Energy Cooperation or Illusion?" World Affairs (Summer 1981):55-82.

Kemezic, Paul. "The Permanent Crisis: Changes in the World Oil System." Orbis 23 (Winter 1980):761-84.

Khot, Nitin. "The Quest for a New Relationship in the World of Oil." Alternatives 5 (January 1980):489-515.

Klinghoffer, Arthur J. "U.S. Foreign Policy and the Soviet Energy Predicament." Orbis (Fall 1981):557-78.

Kuniholm, Bruce R. "What the Saudis Really Want: A Primer for the Reagan Administration," Orbis (Spring 1981):107-21.

Lantzke, Ulf. "The International Energy Agency." Annuaire European 26 (1978):41-65.

Levy, Walter J. "Oil: An Agenda for the 1980s." Foreign Affairs 59 (Summer 1981):1079-1101.

________. "Oil and the Decline of the West." Foreign Affairs 58 (Summer 1980):999-1015.

Lieber, Robert J. "Europe and American in the World Energy Crisis." International Affairs (London) (October 1979):531-45.

Long, David E. "The United States and the Persian Gulf." Current History (January 1979):27-30, 37-38.

Mikdashi, Zuhayr. "Oil Exporting Countries and Oil-Importing Countries: What Kind of Interdependence?" Millennium (Spring 1980):1-20.

Morano, Louis. "Multinationals and Nation-States: the Case of ARAMCO." Orbis (Summer 1979):447-70.

Morse, Edward L. "Energy Security and International Preparedness." Department of State Bulletin (May 1981):34-37.

Nau, Henry R. "Securing Energy." Washington Quarterly (Summer 1981):107-21.

Nutting, Brian. "Congress to Review U.S. Plans for Meeting Energy Shortages: U.S. Preparedness Questioned." Congressional Quarterly Weekly Report (7 January 1984):21-25.

Plant, Steven E. "OPEC Is Not a Cartel." Challenge (November 1981):18-24.

Renner, Michael. "Restructuring the World Energy Industry." MERIP Reports (January 1984):12-17, 25.

Sampson, Martin W., III. "International Policy Coordination: Issues in OPEC and EACM." Monograph Series in World Affairs 19 (1982):3-13.

Schlesinger, James R. "The Geopolitics of Energy." Washington Quarterly 2 (Summer 1979):3-7.

Schuler, G. Henry M. "Coping with Oil Dependence." Washington Quarterly (Winter 1983):53-59.

Scott, Bruce R. "OPEC, the American Scapegoat; Special Report: Mistaken U.S. Policies Have Been a Prime Cause of Rising Oil Prices." Harvard Business Review (January-February 1981):6-30.

Shaw, C. A. "Energy and a New World Economic Order." Year Book of World Affairs 36 (1982): 82-95.

Singer, S. Fred, and Stamas, Stephen. "An End to OPEC?" Foreign Policy (Winter 1981-82):115-25.

Stobaugh, Robert, and Yergin, Daniel. "Energy: An Emergency Telescoped." Foreign Affairs vol. 58, no. 3 (1979):563-95.

Tanzer, Michael, and Zorn, Stephen. "OPEC's Decade: Has It Made a Difference?" MERIP Reports (January 1984):8-11.

Tucker, Robert W. "Oil and American Power--Six Years Later." Commentary 68 (September 1979): 35-42.

Tuma, Elias H. "Strategic Resources and Viable Interdependence: The Case of Middle Eastern Oil." The Middle East Journal (Summer 1979): 269-87.

Willrich, Mason. "A Workable International Nuclear Energy Regime." Washington Quarterly 2 (Spring 1979):13-30.

## Government Documents

U.S. Congress. House of Representatives. Committee on Energy and Commerce. Energy Emergency Preparedness Policy: The Unspoken Agenda. Report. Washington, D.C.: Government Printing Office, 1984.

U.S. House of Representatives. Committee on Energy and Commerce. The United States in a Changing World Economy: The Case for an Integrated Domestic and International Commercial Policy. Washington, D.C.: Government Printing Office, 1983.

U.S. House of Representatives. Committee on Energy and Commerce. Subcommittee on Fossil and Synthetic Fuels. DOE's Emergency Preparedness Budget. Hearings. Washington, D.C.: Government Printing Office, 1982.

U.S. House of Representatives. Committee on Government Operations. Environment, Energy and Natural Resources Subcommittee. Effect of Iraqi-Iranian Conflict on U.S. Energy Policy. Hearing. Washington, D.C.: Government Printing Office, 1981.

U.S. House of Representatives. Committee on Government Operations. Subcommittee on Environment, Energy and Natural Resources. Alternatives to Dealing with OPEC. Hearings. Washington, D.C.: Government Printing Office, 1979.

U.S. House of Representatives. Committee on Government Operations. Subcommittee on Environment, Energy and Natural Resources. Impact on U.S. Energy Policy of Recent OPEC Activities. Hearings. Washington, D.C.: Government Printing Office, 1983.

U.S. House of Representatives. Committee on Interstate and Foreign Commerce. Subcommittee on Energy and Power. Oil Imports: A Range of Policy Options. Report. Washington, D.C.: Government Printing Office, 1979.

U.S. Congress. Joint Economic Committee. Multinational Oil Companies and OPEC: Implications for U.S. Policy. Hearings before the Subcommittee on Energy of the Joint Economic Committee, 94th Congress, 2nd session. Washington, D.C.: Government Printing Office, 1977.

U.S. Congress. Joint Economic Committee. Subcommittee on International Trade, Finance, and Security Economics. Energy in Soviet Policy. Report. Washington, D.C.: Government Printing Office 1981.

U.S. Congress. Senate. Committee on Energy and Natural Resources. Energy Emergency Preparedness. Hearings. Washington, D.C.: Government Printing Office, 1983.

U.S. Congress. Senate. Committee on Energy and Natural Resources. Energy Emergency Preparedness Act Amendments of 1983; and the International Energy Agency Program. Hearing. Washington, D.C.: Government Printing Office, 1983.

U.S. Congress. Senate. Committee on Energy and Natural Resources. Geopolitics of Oil. Hearings. Washington, D.C.: Government Printing Office, 1980. 2pts.

U.S. Congress. Senate. Committee on Energy and Natural Resources. World Petroleum Outlook, 1983. Hearing. Washington, D.C.: Government Printing Office, 1983.

U.S. Congressional Budget Office. The World Oil Market in the 1980s: Implications for the United States. Washington, D.C.: CBO, 1980.

U.S. General Accounting Office. The Economic and Energy Effects of Alternative Oil Import Policies. Report to the Congress by the Comptroller General of the United States. Washington, D.C.: GAD, 1979.

________. The United States and International Energy Issues. Report to the Congress by the Comptroller General of the United States. Washington, D.C.: GAD, 1978.

U.S. Office of Technology Assessment. World Petroleum Availability, 1980-2000. Technical Memorandum. Washington, D.C.: Government Printing Office, 1980.

# FOOD

## Books

Aiken, William, and LaFollette, Hugh. World Hunger and Moral Obligation. Englewood Cliffs, NJ: Printice-Hall, 1977.

Balaam, David N., ed. Food Politics. Montclair, NJ: Allanheld, 1981.

Ball, Nicole. World Hunger: A Guide to the Economic and Political Dimensions. Santa Barbara, CA: ABC-Clio, 1981.

Brown, Lester R. Our Daily Bread. New York: Foreign Policy Association, 1975.

________. The Politics and Responsibility of the North American Breadbasket. Washington, D.C.: Worldwatch Institute, 1975.

________. Redefining National Security. Washington, D.C.: Worldwatch Institute, 1977.

Cathie, John. The Political Economy of Food Aid. New York: St. Martin's Press, 1982.

Cline, William R. Policy Alternatives for a New Economic Order. New York: Praeger, 1979.

Cochrane, Willard W. Feast or Famine. Washington, D.C.: National Planning Association, 1974.

Cohn, Theodore H. Canadian Food Aid: Domestic and Foreign Policy Implications. Denver: University of Denver, 1979.

Commission on Critical Choices for Americans. Vital Resources. Lexington, MA: Lexington Books, 1977.

Committee for Economic Development. A New U.S. Farm Policy for Changing World Food Needs. New York: Committee for Economic Development, 1974.

Destler, I.M. Making Foreign Economic Policy. Washington, D.C.: Brookings Institution, 1980.

Food Policy and Farm Programs. New York: Academy of Political Science, 1982.

The Future of American Agriculture as a Strategic Resource. Washington, D.C.: Conservation Foundation, 1980.

Hansen, Roger D. The U.S. and World Development. New York: Praeger, 1976.

Hopkins, Raymond F. Food in the Global Arena. New York: Holt, Rinehart, and Winston, 1982.

Hopkins, Raymond F., and Puchala, Donald J. Global Food Interdependence. New York: Columbia University Press, 1980.

Hopkins, Raymond F., Puchala, Donald J., and Talbot, Ross B., eds. Politics and Agricultural Development: Case Studies in the Public Policy of Rural Modernization. Boulder, CO: Westview Press, 1979.

Hough, Richard L. Economic Assistance and Security: Rethinking U.S. Policy. Washington, D.C.: National Defense University Press, 1982.

Johnson, David G. U.S. Agriculture in a World Context. New York: Praeger, 1974.

Johnson, Gale. Food and Agricultural Policy for the 1980s. Washington, D.C.: American Enterprise Institute, 1981.

Knutson, Ronald D., Perm, J.B., and Boehm, William T. Agricultural and Food Policy. Englewood Cliffs, NJ: Prentice-hall, 1983.

Lappe, Frances Moore, Aid as Obstacle. San Francisco: Institute for Food and Development Policy, 1980.

Linowitz, Sol M. World Hunger: A Challenge to American Policy. New York: Foreign Policy Association, 1980.

Lucas, George R. Lifeboat Ethics. New York: Harper & Row, 1976.

Lynch, James A., and Tasch, Edward B. Food Production and Public Policy in Developing Countries: Case Studies. New York: Praeger, 1983.

Morss, Elliott R. U.S. Foreign Aid. Boulder, CO: Westview Press, 1982.

Nelson, Jack A. Hunger for Justice: The Politics of Food and Faith. Maryknoll, NY: Orbis Books, 1980.

Power, Jonathan. World of Hunger. London: Temple Smith, 1976.

The Role of U.S. Agriculture in Foreign Policy. New York: Praeger, 1979.

Schapsmeier, Edward L. Ezra Taft Benson and the Politics of Agriculture. Dansville, IL: Interstate Printers & Publishers, 1975.

Schneider, William. Food, Foreign Policy, and Raw Materials Cartels. New York: National Strategy Information Center, 1976.

Schramm, Carl J. United States Strategies for World Hunger. Toronto: Canadian Institute of International Affairs, 1976.

Sokol, John F. Food--Our Next Crisis. Toronto: Canadian Institute of International Affairs, 1977.

Stevens, Christopher. Food Aid and the Developing World. New York: St. Martin's Press, 1979.

Tarrant, John R. Food Policies. New York: Wiley, 1980.

Timmer, Peter C., Falcon, Walter P., and Pearson, Scott R. Food Policy Analysis. Baltimore: Johns Hopkins University Press, 1983.

Wallerstein, Mitchel B. Food for War-Food for Peace. Cambridge, MA: MIT Press, 1980.

Weinbaum, Margin G. Food, Development, and Politics in the Middle East. Boulder, CO: Westview Press, 1982.

Weiss, Thomas G., and Jordan, Robert S. The World Food Conference and Global Problem Solving. New York: Praeger, 1976.

Wilson, Graham K. Special Interests and Policymaking: Agricultural Policies and Politics. New York: Wiley, 1977.

## Journals

Avery, Dennis T. "International Commodity Agreements." Department of State Bulletin (November 1981):34-43.

Bartholomew, Wayne, and Wing, George A. "Arab Petroleum-American Food." Business Horizons (December 1974):5-14.

Butler, Nicholas. "The Ploughshares War between Europe and America." Foreign Affairs (Fall 1983):105-22.

________. "The U.S. Grain Weapon." The World Today (February 1983):52-59.

Cohn, Theodore. "The 1978-79 Negotiations for an International Wheat Agreement: An Opportunity Lost?" International Journal 35 (Winter 1979-80):132-49.

"Controversy over Foreign Policy Export Controls." Congressional Digest (June-July 1983):161-92.

Dam, Kenneth W. "U.S. Foreign Policy and Agricultural Trade." Department of State Bulletin (February 1984):19-22.

Feld, Werner J. "Implementation of the European Community's Common Agricultural Policy: Expectations, Fears, Failures." International Organization 33 (Summer 1979):335-63.

"Food: Potent U.S. Weapon: Interview with Earl L. Butz." U.S. News & World Report (16 February 1976):26-28.

Fry, James. "Problems Surrounding the Control of a Common Fund for Commodities." Millenium 7 (Winter 1978-79):211-21.

Gilmore, Richard. "Grain in the Bank." Foreign Policy 38 (Spring 1980):168-81.

Hillman, Jimmye S. "U.S. and EC Agricultural Trade Policies." Intereconomics (March/April 1983): 72-77.

Hormats, Robert D. "Agricultural Trade with the European Community." Department of State Bulletin (March 1982):43-45.

"Hot Debate: What U.S. Owes to the World's Hungry." U.S. News & World Report (16 December 1974): 20-22.

Paarlberg, Robert L. "Food as an Instrument of Foreign Policy." Proceedings of the Academy of Political Science 34 (1982):25-39.

Patterson, Kathleen. "Keeping Them Happy Down on the Farm." Foreign Policy 36 (Fall 1979):63-70.

"The Question of the U.S. Role in Allocation of Scarce World Resources." Congressional Digest (August/September 1975):193-224.

Riggs, Robert E. "The FAO and the USDA: Implications for Functionalist Learning." Western Political Quarterly 33 (September 1980):314-29.

Rothschild, Emma. "Food Politics." Foreign Affairs (January 1976):285-307.

Salisbury, Grant. "Food as a Weapon-Will U.S. Ever Use It? Oil is a Powerful Lever; Why Not Food?" U.S. News & World Report (2 June 1984): 50-51.

Schubert, James N. "The Impact of Food Aid on World Malnutrition." International Organization 35 (Spring 1981):329-54.

Sheets, Hal. "Big Money in Hunger. A Searching Look at Who Really Controls U.S. Food Policy." Worldview (March 1975):10-15.

Shefrin, Frank. "The Agriculture Agencies: Objectives and Performance." International Journal 35 (Spring 1980):263-91.

Sorenson, David S. "Food for Peace or Defense and Profit? The Role of P.L. 480, 1963-73." Social Science Quarterly 60 (June 1979):62-71.

Talbot, Ross B. "The International Fund for Agricultural Development." Political Science Quarterly 95 (Summer 1980):261-75.

Talbot, Ross R. [Food Policies and Institutions] Review. Orbis 24 (Winter 1981):893-900.

## Government Documents

Law of International Water Resources. Rome: Food and Agricultural Organization of the United Nations, 1980.

United Nations. Committee of the Whole. Food and Agriculture in Global Perspective. New York: Pergamon, 1980.

U.S. Congress. House of Representatives. Committee on Foreign Affairs. World Food Situation.

Hearings. Washington, D.C.: Government Printing Office, 1983.

U.S. Congress. House of Representatives. Joint Committee on Economics. Toward the Next Generation of Farm Policy. Document. Washington, D.C.: Government Printing Office, 1983.

U.S. Congress. Senate. Committee on Foreign Relations. Rapid Population Growth in Third World Countries. Report. Washington, D.C.: Government Printing Office, 1982.

United States Presidential Commissioner on World Hunger. Overcoming World Hunger. Washington, D.C.: Government Printing Office, 1980.

# MARINE MINERALS

## Books

Akinsanya, Adeoye A. The Law of the Sea: Unilateralism or Multilateralism? Lagos, Nigeria: University of Lagos Press, 1980.

Alexander, Lewis M. Regional Arrangements in Ocean Affairs. Washington, D.C.: U.S. Office of Naval Research, Fleet Analysis and Support Division, 1977.

Amacher, Ryan C., and Sweeney, Richard James, eds. The Law of the Sea, U.S. Interests and Alternatives. Washington, D.C.: American Enterprise Institute for Public Policy Research, 1976.

Anand, Ram Prakash. Origin and Development of the Law of the Sea: History of International Law Revisited. Boston, MA: Martinus Nijhoff, 1983.

Armstrong, John M., and Ryner, Peter C. Ocean Management: A New Perspective. Ann Arbor, MI: Ann Arbor Science, 1981.

Baram, Michael, et al. Marine Mining of the Continental Shelf: Legal, Technical, and Environmental Considerations. Cambridge, MA: Ballinger, 1978.

Barkenbus, Jack N. Deep Seabed Resources: Politics and Technology. New York: Free Press, 1979.

Brewer, William C., Jr. Transfer of Mining Technology to the International Enterprise. Charlottesville, VA: Michie, 1980.

Buzan, Barry. Seabed Politics. New York: Praeger, 1976.

Deepsea Mining: Selected Papers from a Series of Seminars Held at the Massachusetts Institute of Technology in December 1978 and January 1979. Cambridge, MA: MIT Press, 1980.

Eckert, Ross D. The Enclosure of Ocean Resources: Economies and the Law of the Sea. Stanford: Stanford University, Hoover Institution Press, 1979.

Eskin, Otho E. Law of the Sea and the Management of Multilateral Diplomacy. Charlottesville, VA: Michie, 1979.

Friedheim, Robert L., ed. Managing Ocean Resources: A Primer. Boulder, CO: Westview Press, 1979.

Galey, Margaret E. Marine Environmental Affairs Bibliography. Kingston: Law of the Sea Institute, University of Rhode Island, 1977.

Jarry, John R. Deep Seabed Mining in the Law of the Sea Negotiation (I): The Contours of a Compromise. Charlottesville, VA: Michie, 1978.

Hauser, Wolfgang. The Legal Regime for Deep Seabed Mining Under the Law of the Sea Convention. Deventer, The Netherlands: Kluwer, 1983.

Hollock, Ann L. U.S. Foreign Policy and the Law of the Sea. Princeton, NJ: Princeton University Press, 1981.

Hoole, Francis W., Friedheim, Robert L., and Hennessey, Timothy M., eds. Making Ocean Policy: The Politics of Government Organization and Management. Boulder, CO: Westview Press, 1981.

International Law of the Sea and the Future of Deep Seabed Mining: Proceedings of the John Bassett Moore Society of International Law Symposium and the American Society of International Law Regional Meeting, Charlottesville, Virginia, November 16, 1974. Charlottesville, VA: Accent Printing, 1975.

Juda, Lawrence. The United States without the Law of the Sea Treaty: Opportunities and Costs. Wakefield: University of Rhode Island, Ocean Management Studies, 1983.

Lauresne, Finn. Superpower at Sea: U.S. Ocean Power. New York: Praeger, 1983.

Luard, Evan. The Control of the Sea-Bed: An Updated Report. New York: Taplinger Publishing, 1977.

Mangone, Gerard J., ed. American Strategic Minerals. New York: Crane Russak, 1984.

________. Marine Policy for America: The United States at Sea. Lexington, MA: Lexington Books, 1977.

Marine Minerals: An Alternative Mineral Supply: National Ocean Goals and Objectives for the 1980s. Washington, D.C.: National Advisory Committee on Oceans and Atmosphere, 1983.

Maxwell, Richard C., and Robertson, Horace B., eds. The Law of the Sea: Where Now? Durham, NC: School of Law, Duke University, 1983.

Moore, John N., et al. Deep Seabed Mining in the Law of the Sea Negotiation (II): Toward a Balanced Development System. Charlottesville, VA: Michie, 1979.

Murphy, John M., et al. Managing National Ocean Resources: Papers. Charlottesville, VA: Michie, 1980.

Oda, Shigeru. International Law of the Resources of the Sea. Germantown, MD: Sijthoff, 1979.

Ogley, Roderick. Internationalizing the Seabed. Brookfield, VT: Gower, 1984.

Oxman, Bernard H., Caron, David D., and Buderi, Charles L. O., eds. Law of the Sea: U.S. Policy Dilemma. San Francisco: Institute for Contemporary Studies, 1983.

Papadakis, Nikos. International Law of the Sea: A Bibliography. Germantown, MA: Sijthoff and Noordhoff, 1980.

Post, Alexandra Merle. Deepsea Mining and the Law of the Sea. Boston, MA: M. Nijhoff Publishers, 1983.

Sebernius, James K. Negotiating the Law of the Sea. Cambridge, MA: Harvard University Press, 1984.

Shusterich, Kurt M. Resource Management and the Oceans: The Political Economy of Deep Seabed Mining. Boulder, CO: Westview Press, 1982.

Simmonds, Kenneth R., ed. New Directions in the Law of the Sea. New York: Oceana Publications, 1983.

Walsh, Don. The Law of the Sea: Issues in Ocean Resource Management. New York: Praeger, 1977.

## Journals

Adebe, A. O. "The Group of 77 and the Establishment of the International Seabed Authority." Ocean Development and International Law 7 (1979):31-64.

Agrait, Luis E. "The Third United Nations Conference on the Law of the Sea and Non-Independent States." Ocean Development and International Law 7 (1979):19-30.

Alexander, Tom. "The Reaganites' Misadventure at Sea." Fortune (23 August 1982):129-44.

Allott, Philip. "Power Sharing in the Law of the Sea." American Journal of International Law 77 (January 1983):1-30.

Al-Mour, Awadh M. "The Legal Status of the Exclusive Economic Zone." Revue Egyptienne de Droit International 33 (1977):35-69.

Anderson, John W. "The Law of the Sea: From Grotius to Pardo." Southeasten Political Review 10 (Spring 1982):89-116.

Arnold, Dennis W. "The Customary Norm and the Deep Seabed." Ocean Development and International Law 9 (1981):1-59.

Arnold, Frederick. "Toward a Principled Approach to the Distribution of Global Wealth: An Impartial Solution to the Dispute over Seabed Manganese Nodules." San Diego Law Review 17, no. 3 (1980).

Arnold, Dennis W. "The Customary Norm and the Deep Seabed." Ocean Development and International Law 9 (1981):1-59.

Bailey, J. S. "Future of the Exploitation of the Resources of the Deep Seabed and Subsoil." Law and Contemporary Problems 46 (Spring 1983):71-93.

Bandow, Doug. "The United States versus the Law of the Sea." Journal of Social, Political and Economic Studies 7 (Fall 1982):195-205.

Barston, R. P. "The Law of the Sea. The Conference and After." Journal of World Trade Law 17 (May-June 1983):207-23.

Bennpouna, Mohamed. "Les Droits d'Exploitation des Ressources Minerals des Oceans" [Exploitation Rights for Ocean Mineral Resources]. Revue Generale de Droit International Public 84 (1980):120-43.

Bleecher, M. D. "Equitable Delimitation of Continental Shelf." American Journal of International Law 73 (January 1979):61-88.

Borgese, Elisabeth Mann. "Law of the Sea: The Next Phase." Third World Quarterly 4 (October 1982):698-718.

Brewer, William C., Jr. "Deep Seabed Mining: Can an Acceptable Regime Ever be Found?" Ocean Development and International Law 11 (1982): 25-67.

Brown, E. D. "Seabed Mining: From UNCLOS to Prep Com." Marine Policy (April 1984):151-64.

Bulmer, Charles. "The Third United Nations Conference on the Law of the Sea: Some Problems in the Seabed Negotiations." World Affairs 141 (Spring 1979):337-50.

Burke, W. Scott, and Brokau, Frank S. "Law at Sea." Policy Review 20 (Spring 1982):71-82.

Burke, William T. "National Legislation on Ocean Authority Zones and the Contemporary Law of the Sea." Ocean Development and International Law 9 (1981):289-22.

Caron, David D. "Municipal Legislation for Exploitation of the Deep Seabed." Ocean Development and International Law vol. 8, no. 4 (1980): 259-97.

Charney, Jonathan L. "The Offshore Jurisdiction of the States of the United States and the Provinces of Canada." Ocean Development and International Law 12 (1983):301-35.

Curtis, Clif. "In Everyone's Interest: There Must Be a Law for the Sea." Oceans (January 1982): 2-3.

D'Amato, Anthony. "An Alternative to the Law of the Sea Convention." American Journal of International Law (April 1983):281-85.

Dickson, David. "Scuttling the Sea-Law Treaty." _The Nation_ (30 May 1981):649, 665-68.

Evriviades, Euripides L. "The Third World's Approach to the Deep Seabed." _Ocean Development and International Law_ 11 (1982):201-64.

Flanagan, Joseph P. "Manganese Nodules: A Huge Resource for the Future." _Sea Technology_ (August 1984):18-21.

Gamble, John King, Jr. "Where Trends the Law of the Sea?" _Ocean Development and International Law_ 10 (1981):61-91.

Gopalakrishnan, Chennat. "Multinational Corporations, Nation-States and Ocean Resource Management: The Impact of the World's 200-mile Economic Zone on Multinational and National Development." _American Journal of Economics and Sociology_ 28 (July 1979):253-70.

Grolin, Jesper. "The Future of the Law of the Sea: Consequences of a Non-Treaty or Non-Universal Treaty Situation." _Ocean Development and International Law_ 13 (1983):1-31.

Hegelsom, G. J. van. "The Argument for an Interim Arrangement." _Marine Policy_ (July 1981):260-63.

Irwin, Paul C. "Settlement of Maritime Boundary Disputes: An Analysis of the Law of the Sea Negotiations." _Ocean Development and International Law_ 8 (1980):105-48.

Jones, William B. "The International Sea-bed Authority without U.S. Participation." _Ocean Development and International Law_ 12 (1983): 151-71.

Kimball, Lee. "Regional Marine Resources Development: Growth by Necessity," _Ocean Yearbook_ 3:157-97.

Kimball, Lee, and Schneider, Adolf R. H. "Cracking the Hard Cores--Highlighting the Differences. UNCLOS III Now on the Way to Caracas? _Environmental Policy and Law_ 5 (June 1979):128-37.

King, Alwyn H. "The Strategic Minerals Problem: Our Domestic Options." _Parameters_ (September 1982):44-52.

Koch, Kathy. "Deep Seabed Mining Passed by House." _Congressional Quarterly Weekly Report_ (14 June 1980):1624-25.

Laursen, Finn. "Security Versus Access to Resources: Explaining a Decade of U.S. Ocean Policy." World Politics (January 1982):197-229.

"The Law of the Sea--Where Now?" Law and Contemporary Problems (Spring 1983):entire issue.

Lehman, John. "The Law of the Sea." Defense 82 (March 1982):1-9.

Lowe, A. V. "The Development of the Concept of the Contiguous Zone." British Yearbook of International Law 52 (1981):109-69.

Malone, James L. "Law of the Sea and Oceans Policy." Department of State Bulletin (October 1982):48-50.

________. "Who Needs the Sea Treaty?" Foreign Policy (Spring 1984):44-63.

Manley, Robert H. "Developing Nation Imperatives for a New Law of the Sea: UNCLOS I and III as Stages in the International Policy Process." Ocean Development and International Law (1979):9-18.

Martin, Jochen. "Deep-Sea Mining between Convention and National Legislation." Ocean Development and International Law 10, no. 1-2 (1981):175-85.

Marston, Geoffrey. "The Evolution of the Concept of Sovereignty over the Bed and Subsoil of the Territorial Sea." British Yearbook of International Law 48 (1976-77):321-32.

Mehrish, B. N. "The Role of the Third World on the Law of the Sea Negotiations." Foreign Affairs Reports 28 (March 1979):31-49.

Moore, John Norton. "Charting a New Course in the Law of the Sea Negotiations." Denver Journal of International Law and Policy 10 (Winter 1981):207-19.

Moss, Ronald S. "Insuring Unilaterally Licensed Deep Seabed Mining Operations against Adverse Rulings by the International Court of Justice: An Assessment of the Risk." Ocean Development and International Law 14, no. 2 (1984):161-91.

Murphy, John M. "The Politics of Manganese Nodules: International Considerations and Domestic Legislation." San Diego Law Review 16, no. 3 (1979):531-54.

Ogley, Roderick C. "The Ocean Revolution." British Journal of International Studies 4 (October 1978):270-81.

Oxman, Bernard H. "The Third United Nations Conference on the Law of the Sea: The Seventh Session" (1978). American Journal of International Law 73 (January 1979):1-41.

Payne, Richard J. "U.S. Foreign Policy at Sea: National Security on the Seabed." World Today (October 1983):393-99.

Pirtle, Charles E. "Alternative Regimes for Harvesting the Seabed: A Review Article." Ocean Development and International law 9, no. 1-2 (1981):77-99.

Pollard, Duke, E. "The International Seabed Authority: Implications for Transnational Corporate Power." Caribbean Yearbook of International Relations (1977):83-114.

Post, Alexandra M. "United Nations Involvement in Ocean Mining." Ocean Development and International Law 10 (1982):275-313.

Puri, Rama. "Exclusive Economic Zone: A New Dimension in the Law of the Sea." Indian Political Science Review 14 (January 1980):39-54.

________. "If UNCLOS III Fails . . ." IDSA Review 10 (Spring 1982):424-40.

Ratiner, Leigh S. "The Law of the Sea: A Crossroads for American Foreign Policy." Foreign Affairs 60 (summer 1982):10006-21.

Richardson, Elliot L. "Power, Mobility and the Law of the Sea." Foreign Affairs 54 (Spring 1980):902-19.

________. "The Politics of the Law of the Sea." Ocean Development and International Law 11 (1982):9-23.

________. "Seabed Mining and the Law of the Sea." Department of State Bulletin (December 1980): 60-64.

"The Sea Bed International Area." Impact of Science on Society 33, no. 3-4 (1983):327-34.

Silva, Neville de. "The Sea Minus the U.S." The World Press Review (July 1982):55.

Stavridis, James. "Naval Strategy and National Ocean Policy." U.S. Naval Institute Proceedings (July 1984):43-47.

Swing, John T. "Law of the Sea." Bulletin of the Atomic Scientists (May 1983):14-19.

Turbyville, Linda. "Plumbing the Depths." Sea Power (January 1981):23-28.

Van Dyke, Jon M., and Teichmann, David L. "Transfer of Seabed Mining Technology." Ocean Development and International Law 13, no. 4 (1984):427-55.

Watt, Donald C. "Law of the Sea Conference and the Deep Seabed Mining Issue: the Need for an Agreement." International Affairs (Winter 1981-82):78-94.

________. "Toward a New Order for the World's Oceans?" Round Table (January 1983):25-33.

Willsey, F. P. "The Deep Seabed Hard Mineral Resources Act and the Third United Nations Conference on the Law of the Sea: Can the Conference Meet the Mandate Embodied in the Act?" San Diego Law Review 18, no. 3 (1981): 509-32.

## Government Documents

United Nations. Dag Hammarskjold Library. The Sea: A Select Bibliography on the Legal, Political, Economic and Technological Aspects, 1978-1979. New York: United Nations, 1980.

U.S. Congress. House of Representatives. Committee on Foreign Affairs. The Status of the Third United Nations Conference on the Law of the Sea, Spring 1980. Hearing. Washington, D.C.: Government Printing Office, 1980.

U.S. Congress. House of Representatives. Committee on Foreign Affairs. U.S. Policy and the Third United Nations Conference on the Law of the Sea. Hearings. Washington, D.C.: Government Printing Office, 1981.

U.S. Congress. House of Representatives. Committee on Foreign Affairs. Subcommittee on International Organizations. Deep Seabed Mining and the Law of the Sea. Hearings. Washington, D.C.: Government Printing Office, 1977.

U.S. Congress. House of Representatives. Committee on Foreign Affairs. Congress and Foreign Policy: 1980. Report. Washington, D.C.: Government Printing Office, 1981.

U.S. Congress. House of Representatives. Committee on Foreign Affairs. U.S. Foreign Policy and the Law of the Sea. Hearings. Washington, D.C.: Government Printing Office, 1982.

U.S. Congress. Senate. Committee on Foreign Relations. Subcommittee on Arms Control, Oceans, International Operations, and Environment. Law of the Sea Negotiations. Hearing. Washington, D.C.: Government Printing Office, 1982.

U.S. Department of Commerce. National Oceanic and Atmospheric Administration. Office of Coastal Zone Management. Coastal Zone Management in the 80s. Washington, D.C.: Government Printing Office, 1981.

U.S. Department of State. Bureau of Public Affairs. U.S. Exclusive Economic Zone. Washington, D.C.: Government Printing Office, 1983.

U.S. General Accounting Office. Uncertainties Surround Future of U.S. Ocean Mingin. Washington, D.C.: Government Accounting Office, 1983.

________. Impediments to U.S. Involvement in Deep Ocean Mining Can Be Overcome. Washington, D.C.: Government Accounting Office, 1982.

U.S. Laws, Statutes, etc. "Deep Seabed Hard Minerals Resources Act," United States Statutes at Large. Public Law 96-283, 96th Congress, 2d sess. Washington, D.C.: Government Printing Office, 1981.

________. "Deep Seabed Hard Minerals Resources Act, Amendment." United States Statutes at Large. Vol. Public Law 97-416, 97th Congress, 2d sess. Washington, D.C.: Government Printing Office, 1984. vol. 96, pt. 2.

U.S. National Advisory Committee on Oceans and Atmosphere. Marine Minerals: An Alternative Mineral Supply, National Ocean Goals and Objectives for the 1980s. Washington, D.C.: Government Printing Office, 1983.

# INDEX

# ABOUT THE EDITOR AND CONTRIBUTORS

Robert S. Jordan is Distinguished Visiting Professor at the Naval War College, on leave of absence from his appointment as Professor of Political Science at the University of New Orleans. His most recent book (with Werner Feld) is *Europe in the Balance: The Changing Context of European International Politics* (London: Faber and Faber, 1985).

Donald J. Puchala is Director of the Institute of International Studies, University of South Carolina. He is co-author (with Raymond F. Hopkins) of *Global Food Interdependence* (New York: Columbia University Press, 1980), and annual editor of *Issues Before the United Nations General Assembly*.

Joan Edelman Spero is Senior Vice President for International Corporate Affairs of the American Express Co. She has served as United States Representative to UNESCO, and is author of *The Politics of International Economic Relations*, (New York: St. Martin's Press, 1981).

Janet K. Schmidt teaches in the Stanford Program in Values, Technology, Science, and Society, at Stanford University.

Roger A. Coate is Assistant Professor in the Department of Government and International Studies at the University of South Carolina. He is author of *Global Issue Regimes* (New York: Praeger, 1982).

John P. Renninger is a Research Officer in the Division of Research of the United Nations Institute for Training and Research. He is author of *Multinational Cooperation for Development in West Africa* (New York: Pergamon Press, 1979), and various reports in the UNITAR Policy and Efficacy Studies series.

Christopher Herrick is Assistant Professor of Political Science at Muhlenberg College. He formerly was a participant in the South Carolina SeaGrant Consortium, during which he authored *The Evolution of the Fishery Conservation Zone as an Instrument of United States Foreign and Domestic Policy*, and co-authored (with Robert S. Jordan)

Management of an Expanded Territorial Sea: Impact of the Third United Nations Law of the Sea Conference on South Carolina and the Nation.

Richard J. Kessler is a Fellow in the Energy Strategic and Resources Program of the Center for Strategic and International Studies of Georgetown University. He is co-author of the Critical Link: Energy and National Security in the 1980s (Cambridge, Mass.: Ballinger Press, 1984.

Raymond F. Hopkins is Professor of Political Science at Swarthmore College. He is co-author (with Donald J. Puchala) of Global Food Interdependence (New York: Columbia University Press, 1980).